Key Topics in Brain Research

Edited by

A. Carlsson, P. Riederer, H. Beckmann, T. Nagatsu,
S. Gershon, and K. A. Jellinger

Springer-Verlag Wien GmbH

N. Müller (ed.)

Psychiatry, Psychoimmunology, and Viruses

Springer-Verlag Wien GmbH

PD Dr. Dipl.-Psych. Norbert Müller
Psychiatrische Klinik und Poliklinik
Ludwig-Maximilians-Universität München
Nußbaumstraße 7, D-80336 München

Originally published by Springer-Verlag/Wien in 1999

Typesetting: Best-Set Typesetter Ltd., Hongkong

Graphic design: Ecke Bonk
Printed on acid-free and chlorine-free bleached paper
SPIN: 10704402

With 18 Figures

ISSN 0934-1420
ISBN 978-3-211-83249-3 ISBN 978-3-7091-6404-4 (eBook)
DOI 10.1007/978-3-7091-6404-4

Preface

This volume contains the contributions to the "3rd Expert Meeting in Psychoneuroimmunology", held in Munich, Germany in autumn 1997. Clinical psychiatrists, neuroscientists and researchers in basic immunology from Europe and the USA gather together annually at this new traditional meeting to discuss new findings and trends in the relationship between psychiatric disorders and the immune system. Pinpointed at the Munich meeting in particular were the possible influence of viral infections on psychiatric disorders, especially schizophrenia, and the possible pathophysiological mechanisms affecting psychiatric disorders. These mechanisms include viral-induced alteration, cytokine production, and the role of the HLA system.

One chapter is devoted to the influence of viral infections during pregnancy; another to the first results of experimental therapy especially regarding the immune system. The immunological aspects not only of schizophrenia but also of depression, anxiety disorders, sleep and headache were also discussed at the meeting, and two further chapters cover methodological aspects related to clinical and experimental pitfalls in psychoneuroimmunology.

The psychoneuroimmunology of psychiatric disorders is a rapidly expanding field, and research is increasing on the therapeutic aspects. This is reflected by one of the main themes of the "4th Expert Meeting in Psychoneuroimmunology", jointly held in autumn 1998 in Bethesda, Maryland, USA, with the Stanley Foundation Meeting in Neurovirology, devoted to therapeutic strategies deriving from psychoneuroimmunological findings. The "5th Expert Meeting in Psychoneuroimmunology" will be held in 1999 in Innsbruck, Austria.

Munich, Spring 1999 NORBERT MÜLLER

Contents

Immunogenetic aspects of viral or other environmental factors in schizophrenia

R. Wank

Institut für Immunologie, Universität München, Federal Republic of Germany

Summary

Neuronal regulation appears to be vulnerable to cytokines of the immune system. Environmental microbial factors can stimulate the production of such cytokines. The type of cytokine, elaborated in the course of acute and chronic infections, depends on the individual immunogenetic disposition. Examples of immune reponses will be discussed that can lead to a polarized overproduction of some cytokines in dependance of the immunogenetic disposition.

Introduction

The essential contribution of genetic factors to the pathogenesis of schizophrenia has been generally accepted [1–3]. The concordance rate in monozygotic twins of approximately fifty percent suggests a contribution of both genetic and environmental microbial factors. Recently, the search for such environmental factors has focussed on the role of viruses (reviewed in [4], and this volume).

The purpose of this article is not to discuss arguments for or against a certain virus that may or may not be involved in the schizophrenic syndrome. I want to raise awareness for the information that could be gained from inclusion of immunogenetic results in the analysis of the schizophrenic syndrome.

Distorted cytokine and cytokine receptor concentrations and the autoimmune hypothesis of schizophrenia: central or peripheral origin of distortion?

A medieval painting by Hieronymos Bosch depicts the extraction of the stone of folly from the brain of a quietly sitting individual [5].

Where shall we search the stone of schizophrenia? Do we get fooled by searching pathognomonic alterations in the brain of schizophrenics? Certainly, there are slightly enlarged ventricles in post mortem brains of schizophrenics but the brilliant painter Adolph Menzel, who died with a brain weight of 980 grams due to a hydrocephalus, had no sign of mental confusion or abnormal behaviour until his death. Similarly disappointing in general was the fine inspection of the brain with the sensitive tool of the polymerase chain reaction (PCR) in order to identify viral agents, although reports showed elevated antibody titers against the probed viruses [4]. The paucity of findings in the brain could indicate a lack of sophisticated tools or methods, but could also indicate an error of pricipal strategy. If smoke makes the lungs cough we must not find the fire in the lungs. The hypothesis that autoimmune processes could be the basis of the schizophrenic syndrome would also agree with a location of disease processes in the periphery. The hypothesis of an autoimmune process would also agree with decreased concentrations of interleukin 2 (IL-2) in the peripheral blood and increased concentrations of IL-2 in the crebrospinal fluid [5–7]. The importance of these changes in IL-2 concentrations for the schizophrenic disease process is illustrated by the observation that relapses after neuroleptica treatment are accompanied by increased concentrations of IL-2 in the crebrospinal fluid [8].

We have to keep in mind that changes in concentrations of IL-2 or other interleukines like IL-6 as well as changes in IL-2 receptor concentrations can be interpreted as indications for infectious disease processes but not as proofs. Furthermore, the distorted cytokine concentrations must not cause themselves schizophrenic symptoms and could represent simply indicators. For example, other products from infectious diseases like viral peptides could interfere with receptors of transmitters or neurotrophines in the CNS.

However, direct effects of IL-2 on the homeostasis of brain signalling was observed in cancer patients treated with high-dose IL-2 because they displayed symptoms of schizophrenia that ceased with discontinuation of the IL-2 medication [9].

HLA factors and the autoimmune hypothesis of schizophrenia

Several investigators found vulnerability loci for schizophrenia on the short arm of chromosome 6 (reviewed in [10]). This evoked the interest in former and more recent studies of associations between schizophrenia and particular HLA antigens or alleles because chromosome 6p contains also the HLA gene region.

How can different HLA antigens or alleles show significant associations with schizophrenia?

Results of HLA association studies in schizophrenia have met considerable scepticism before the recognition of autoimmune features in schizophrenia. There is still perplexity about the various HLA studies finding different HLA antigens and alleles significantly associated with schizophrenia. One has to inspect three different problem-loaded areas of HLA association studies in order to understand that divergent results of properly conducted HLA association studies must not exclude the involvement of several different HLA antigens or alleles:

a) Common epitopes of HLA antigens or alleles can present the same peptide of viral or other origine, meaning that several allelic HLA molecules could present the same disease inducing peptide. An example would be the antigen HLA-B27 encoded by the seven different HLA class I alleles B27*01-*07. Six of these alleles are significantly associated with ankylosing spondylitis. In this autoimmune disease bacteria especially yersinia, chlamydia and klebsiella are under suspect to deliver the disease peptides.

b) The same single HLA alleles, particularly "promiscuous" HLA class II alleles, can present peptides from different bacteria or virus. An example would be the HLA class II allele DQB1*0602. T cell clones from multiple sclerosis patients carrying the HLA-DQB1*0602 allele were sucessfully restimulated with peptides derived from herpes simplex, helicobacter and other microbes after sensitization to myelic basic protein [11]. Interestingly, the same allele has been found to be associated also with narcolepsy (reviewed in [12]) and with schizophrenia subtypes [13]. This restimulation of T cell clones by peptides from very different microbes illustrates how single T cells can get amplified by different stimuli. The amplified T cells may dominate the T cell response („polarization) by producing a cytokine that in turn can suppress or inactivate other T cells.

c) HLA alleles of different loci can be inherited in linkage disequilibrium, meaning that another than the accused disease-associated allele is the main contributing disease factor.

The last point was often presented in the past before the HLA genes could be analysed molecularly and before the complex biologic functions of HLA genes were recognized. We know now that immune response genes are not linked to HLA genes but are themselves immune response genes. HLA allelic variants of LMP2/7 are involved in processing of viral proteins to peptides, HLA allelic variants of TAP1/TAP2 in selection and transport of these peptides and HLA class I and class II molecules in the presentation of these peptides at the cell surface. So the ability of the individual to respond successfully to viral challenges by proper processing and presenting viral peptides depends to a large degree on both sets of HLA allelic variants inherited by the parents.

Possible modulation of symptoms and strength of association by the alternate HLA haplotype

This mutual influence of alleles of both HLA chromosomal regions (haplotypes) on disease expression has been elaborated mainly in juvenile diabetes [14]. Protectivity-associated HLA haplotypes can modify the effects of disease associated HLA haplotypes. This important aspect has been hardly appreciated in the evaluation of results. For example, rheumatoid arthritis seems to be protective against schizophrenia. Hence, HLA alleles associated with rheumatoid arthritis should be considered as protective factor in the analysis of HLA alleles associated with vulnerability to schizophrenia.

HLA factors, viral infections and cytokine production

The variability of the immune response against HIV has taught us much about the mutual dependence of HLA constellation, immune response and cytokine production. It is also interesting for the theme here that paranoid symptoms occur during the AIDS phase only in ten to twenty percent of the patients. HLA class II alleles are associated with fast and slow progression towards AIDS. The same HLA-class I alleles were reported to show a significant correlation with resistance against HIV in two investigated populations [15]. It would be, most likely, informative to analyse the immunogenetic constellation *and* the cytokine production of HIV infected patients who develop paranoid symptoms.

How do immunogenetic factors influence cytokine production? The best answer we can get in the moment comes from observations in mice. Two mouse strains, BALB/c and C57BL/6 showed a different T-helper 1 development if the cytokine IL-18 (formerly called IFNγ inducer, IFIG) was used for induction but the same, if IL-12 was used [16]. The interest in this potentiation of IL-12 induced Th1 development in BALB/c but not in C57BL/6 mice was raised because BALB/c mice develop IL-4 predominant responses to Leishmania major infection [17].

In this context it would be interesting to know, to which extent the IL-10 like cytokine from EBV [18] can influence the immune response between individuals because of their different immunogenetic dispositions. Infection with EBV is a common event in most populations, but serious illness hits relatively few individuals. Antibody titers against capsid and/or nuclear antigens provide limited information on the gravity of symptoms. Similar problems exist in the evaluation of antibody titers against other microbes. Twenty percent of chlamydia infected individuals do not produce antibodies against the respective chlamydial strain [19].

Conclusions

The same antigen can evoke many distinct immune responses dependend on the presence of immunogenetic factors that differ from individual to individual. This implies also the production of different cytokines upon antigenic stimulation. Polymorphic HLA factors are deeply involved in the immune reponse against environmental factors and influence also cytokine patterns. Individual polymorphic HLA traits should be included therefore in the analysis of autoimmune factors in schizophrenia.

References

1. Kety SS, et al. (1971) Mental illness in the biological and adotpive families of adopted schizophrenics. Am J Psychiatry 128:302–312
2. Gottesmann II (1991) Schizophrenia genesis. WH Freeman and Company, New York
3. Nurnberger JJ, Gershon ES (1992) Genetics. In: Paykell (ed) Handbook of affective disorders. Churchill Livingstone, Edinburgh, pp 131–148
4. Yolken RH, Torrey EF (1995) Viruses, schizophrenia, and bipolar disorder. Clin Microbiol Rev 8:131–145
5. Barondes SH (1993) Molecules and mental illness. Scientific American Library, New York, p V (Preface)
6. Licinio J, Seibyl JP, Altemus M, Charney DS, Krystal JH (1993) Elevated CSF levels of Interleukin-2 in neuroleptic-free schizophrenic patients. Am J Psychiatry 150:9:1408–1410
7. Wilke I, Arolt V, Rothermund M, Weitzsch C, Homberg M, Kirchner H (1996) Investigations of cytokine production in whole blood cultures of paranoid and residual schizophrenic patients. Eur Arch Psychiatry Clin Neurosci 246:279–284
8. Mc Allister CG, et al. (1995) Increases in CSF levels of Interleukin-2 in schizophrenia: effects of recurrence of psychosis and medication status. Am J Psychiatry 152:1291–1297
9. Denicoff KD, et al. (1987) The neuropsychiatric effects of treatment with Interleukin-2 and lymphokine activated killer cells. Ann Intern Med 107:293–300
10. Peltonen L (1995) All out for chromosome six. Nature 378:665–666
11. Wucherpfennig KW, Strominger J (1995) Molecular mimicry in T-cell mediated autoimmunity: viral peptides activate human T-cell clones specific for myelin basic protein. Cell 80:707–718
12. Mignot M, Tafti M, Dement WC, Grumet FC (1995) Narcolepsy and immunity. Adv Neuroimmunol 5:23–27

13. Großkopf A, Müller N, Malo A, Wank R (1998) Potential role for the narcolepsy- and multiple sclerosis-associated HLA allele DQB1*0602 in schizophrenia subtypes. Schizophr Res 30:187–189
14. Segurado OG, Arnaiz-Villena A, Wank R, Schendel DJ (1993) The multifactorial nature of MHC linked susceptibility to insulin-dependent diabetes. Autoimmunity 15:85–89
15. Laumbacher B, Wank R (1998) Recruiting HLA to fight HIV. Nature Med [Vaccine Suppl] 4:5:5
16. Robinson D, et al. (1998) IGIF does not drive Th1 development, but synergizes with IL-12 for IFNγ production, and activates IRAK and NF-κB. Immunity 7:571–581
17. Locksley RM, Scott P (1991) Helper T-cell subsets in mouse Leishmaniasis: induction, expansion and effector function. Immunol Today 12:58–61
18. Hsu DH, et al. (1990) Expression of Interleukine-10 activity by Epstein-Barr virus protein BCRF1. Science 250:830–832
19. Kuo CC, Shor A, Campbell LA, Fukushi H, et al. (1993) Demonstration of Chlamydia pneumoniae in athrosclerotic lesions of coronary arteries. J Infect Dis 167:841–849

Correspondence: Prof. Dr. R. Wank, Institut für Immunologie, Universität München, Goethestrasse 31, D-80336 München, Federal Republic of Germany

The use of subtraction libraries for the identification of RNA species upregulated in the brains of individuals with schizophrenia

R.H. Yolken, N. Johnston, F. Leister, E.F. Torrey,
and **the Stanley Neuropathology Consortium**

The Stanley Neurovirology Laboratory, Johns Hopkins University School of Medicine, Department of Pediatrics, Baltimore, MD, and National Institute of Mental Health, St. Elizabeths, Washington, DC, U.S.A.

Summary

Schizophrenia is a brain disease of unknown etiology. It is likely that individuals with schizophrenia express RNAs within their brain tissues which differ qualitatively and quantitatively from the RNAs expressed by unaffected individuals. We developed a series of brain subtraction libraries which are enriched for RNAs which are differentially expressed by individuals with schizophrenia. Analysis of these libraries identified a number of RNA species associated with brain function and the inflammatory response. In addition, we identified a number of novel RNAs which have not been previously characterized. The identification of brain RNAs which are disease associated might lead to new modalities for the diagnosis and treatment of schizophrenia.

Schizophrenia is a complex neuropsychiatric disorder with worldwide prevalence. While the specific etiopathogenesis of schizophrenia is not known with certainty, numerous studies have indicated that schizophrenia is associated with abnormalities of brain structure and function. Furthermore, numerous studies have identified familial clusters of cases, indicating possible genetic determinants of disease susceptibility [1]. These findings have led to extensive searches for genetic determinants of schizophrenia utilizing linkage analyses and other positional cloning techniques. These analyses have identified genomic regions which display increased rates of inheritance as compared to unaffected controls. However, despite extensive studies of large numbers of individuals and family kindreds, specific genetic defects associated with schizophrenia have not been identified [2,3,4]. This failure has led to the search for environmental factors which might modulate gene expression in the absence of inherited alterations in coding sequences. The role of infection and other environmental factors in the etiology of schizophrenia is supported by a number of studies which have addressed the epidemiology and pathophysiology of this disease [5,6].

Viruses are obligate intracellular parasites which can infect a wide range of host cells. In many cases, viral infection of host cells results in rapid cytolysis with the subsequent release of large numbers of progeny viruses. However, viral genetic material can also integrate into the host genome and alter RNA transcription in the absence of cytopathology. This altered transcription can result in modulation of a range of cellular functions without pathologic evidence of infection or inflammatory response. The effect of viral infection on the host can also be modulated by genetic factors. Genetic determinants of infection include genes which modulate the immune response, the expression of viral receptors, and the susceptibility of cells to soluble factors generated during the course of viral infection [7.8]. It is thus possible that viral infection of neuronal cells can, in combination with genetic determinants of susceptibility, result in the profound alteration of brain function typical of human neuropsychiatric diseases [9].

The characterization of the role of genetic and environmental effects on RNA transcription in neuronal cells requires the unbiased measurement of differentially expressed RNAs in the brains of affected individuals. Recently, molecular biological technologies have been devised which allow for the precise characterization of RNA derived from human brain tissue. These methods are based on the ability of polymerase chain reaction (PCR) to amplify low levels of RNA in a quantitative fashion and to distinguish levels of specific RNAs in cases and controls [10]. Several of these methods involve the "subtraction" of RNAs amplified from individuals with a defined disease process by RNAs derived from matched controls. These procedures result in the generation of pool (or library) or RNA sequences which is enriched for RNA species which are expressed at a different level in affected individuals as compared to unaffected controls. We have applied the method of subtractive hybridization to characterize RNAs which are differentially expressed in the frontal cortex of individuals with schizophrenia.

Methods

Postmortem brain samples were obtained from the Stanley Neuropathology Consortium (details of the consortium and the methods used for brain collection, characterization, and processing can be found at the web site "WWW.Stanleylab.org"). Sections were cut from frozen frontal cortex tissue of 4 individuals with schizophrenia and 4 controls. RNA was extracted as previously described [11] and full length cDNA was prepared by the method of Lukyanov et al. [12] (Fig. 1). The cDNAs from the individuals with schizophrenia were subtracted from cDNAS prepared from the control individuals by

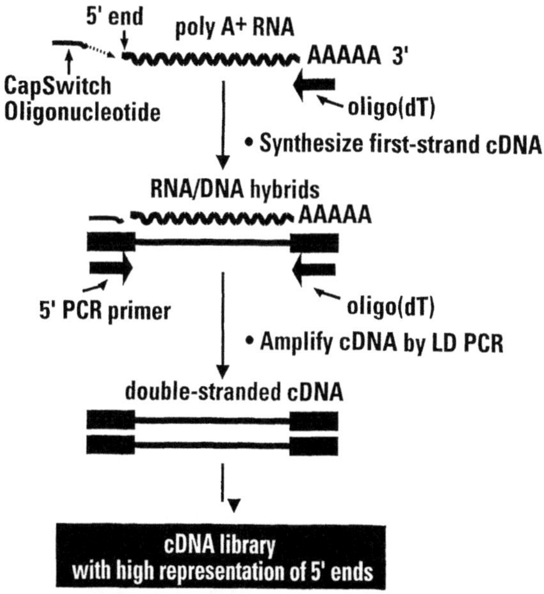

Fig. 1. Outline of method for the preparation of full length cDNA from RNA extracted from brain. The method is described in detail in reference 12 (figure kindly provided by Dr. Luda Diatchenko, Clontech Laboratories, Palo Alto, CA)

the method of Diatchenko et al. [13] (Fig. 2). Reagents for the cDNA generation and subtractions were purchased from Clontech. The cDNA resulting from the subtraction was cloned into the vector pCR2.1 using the T/A Cloning system (Invitrogen). The nucleotide sequences of individual clones were determined using an ABI automated sequencer and the sequences were characterized using the BLAST, UNIGENE, and ORF programs accessed at the National Institutes of Health Web Site "www.ncbi.nlm.nih.gov". Clones were prepared from 2 separate subtraction procedures each employing 2 case and 2 control brains. Clones which consisted of repetitive sequences such as SINE (short interspersed nuclear elements) or LINE (long interspersed nuclear elements) were removed from the analysis.

Results

We analyzed a total of 97 clones from the subtraction libraries generated from the RNA of brains from individuals which schizophrenia. These clones yielded a total of 79 unique sequences. The distribution of these sequences in terms of homologies to sequences contained in available databases is presented in Table 1. We found that 28 (35%) of the sequences, derived from 40 (41%) of the clones displayed identity of high levels of homology with genomic regions that encode genes of known functional significance. An additional 33 (42%) of the sequences, derived from 38 (39%) of the clones displayed high degrees

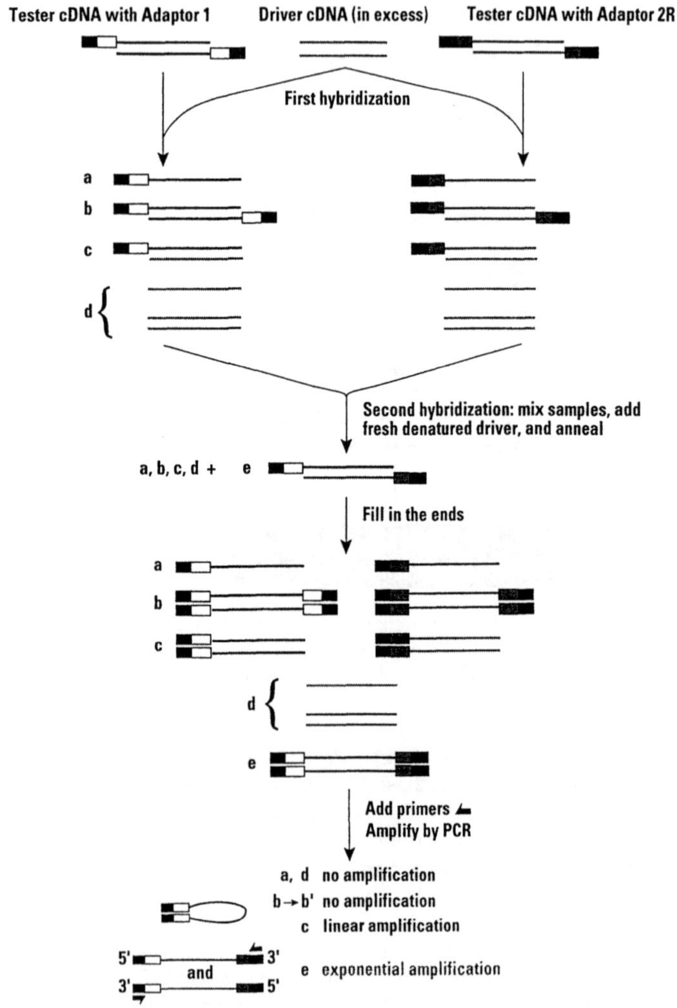

Fig. 2. Outline of the method used for the subtraction of cDNAs. The method is described in detail in [13] (figure kindly provided by Dr. Luda Diatchenko, Clontech Laboratories, Palo Alto, CA)

of homology with sequences derived from human cDNAs for which functions have not been determined. Most of these sequences were derived from expressed sequence tags (ESTs) from brain and other tissue libraries [14,15]. Finally, we found that 18 (23%) of the sequences, derived from 19 (20%) of the clones were did not display significant levels of nucleotide homology with any known human, eukaryotic, or prokaryotic sequences and thus represent novel sequences.

The homologies of the clones that yielded sequences with homologies to genes of known function are presented in Table 2. A total of

Table 1. Databank matches of sequences form subtraction libraries

	No. (%) Clones	No. (%)
Sequences		
Known function	40 (41%)	28 (35%)
Unknown function	38 (39%)	33 (42%)
Novel sequence	19 (20%)	18 (23%)
Total	97	79

Table 2. Sequences with homologies to RNAs encoding proteins of known functional significance

Sequence match	No. matches	Genomic location
X62996/HSU09500/AF03542 Mitochondrial Sequences	7	M
D87666/L08069/D86956 Heat Shock Proteins	4	M
z 47087 RNA polymerase II elongation factor	2	5,17
P34443 ras related GTP binding protein	1	1,10
Q06003 D melanogaster-Probable transcription factor	1	1
P46778 60 s ribosomal protein	1	11
HUMPKC2 protein kinase C beta 1 gene	1	16
AB001106 mRNA for glial maturation factor	1	14
HUMNPM nucleophosmin mRNA	1	5
HSSPMSYN mRNA for spermine synthetase	1	X
D84296 mRNA for TPRIII Downs Syndrome Protein	1	21
HSNCHIM mRNA for n-chimerin	1	M
MMU10355 Mus musculus BALB/c synaptotagenin	1	18
RNU15138 Rattus norvegicus LIC-2 dynein light chain	1	16
D49958 fetus brain mRNA for membrane glycoprotein m6	1	4
X98259 M phase phosphoprotein, mpp8	1	
HUMHNRNA nuclear ribonucleoprotein particle (hnRNP) C protein mRNA	1	12
HSU94319 autoantigen DFS70 mRNA	1	
HUMSCP2B sterol carrier protein 2 mRNA	1	1
KIAA0026 homeobox protein	1	X
HSEAR3 v-erbA related ear-3 gene	1	5
HSPTP1D phosphotyrosine phosphatase. (type 6)	1	6
HSERC55R erc-55	1	15
KIAA0203 neuronal intermediate filiment	1	
HSCALMG2 calmodulin	1	14
D78013 dihydropyrimidinase related protein-2	1	8
HSU29943 Human ELAV-like neuronal protein	1	9
HSITBA4 ITBA4 protein-ring finger protein	1	

Numbers and letter preceding the proteins indicate the Genbank or EMBL ascension numbers. Proteins are human in origin unless otherwise indicated. "M" indicates multiple genetic assignments. Blank spaces indicate that the genomic region is mapped to multiple sites or has not been unambiguously assigned

8 clones had homologies to sequences encoded by the mitochondrial origin. An additional 5 clones displayed sequence homology to heat shock proteins. We also found single clones of sequences homologous to structural proteins within the human or animal central nervous system, such as synaptogenin, chimerin, and neuronal intermediate filament. Additional expressed sequences, such as nucleophosmin, ribonucleoprotein C particle protein, and autoantigen dfs70 are structural proteins which can serve as the targets for antibodies generated as part of an autoimmune response [16,17]. We also identified sequences encoding proteins associated with cellular metabolism. Such sequences include those homologous to protein kinase C, RNA polymerase II elongation factor, M phase phosphoprotein mpp8, *ras* related GTP binding protein and phosphotyrosine phosphatase. Genes encoding these proteins are localized to a large number of human chromosomes.

The homologies to human sequences of unknown function (expressed sequence tags or ESTs) are presented in Table 3. The table also presents the chromosomal location of the ESTs which have been mapped to specific locations. It is of note that multiple expressed sequence tags had been mapped to chromosomes 1, 6, 8, 11 and the X chromosome. All of these chromosomes contain regions that have been associated with susceptibility to schizophrenia [18–22]. Multiple ESTs were also mapped to chromosome 21, which has not yet been associated with susceptibility to schizophrenia but which has been associated with Down's syndrome. There were also single ESTs mapped to additional regions of the human genome as well as ones that have not as yet been localized to discrete genomic regions.

We identified a total of 18 sequences that did not display any significant homology to known human or microbial sequences (Table 4). Several of these sequences have been entered into Genbank with ascension numbers AF017086-AF01795, the remainder are available from the authors pending their addition to the Genbank database. Seven (38.9%) of these sequences, although they did not match known sequences at the nucleotide level, possessed open reading frames with the potential for encoding proteins with homology to viral proteins. These predicated homologous proteins included components of human herpesvirus type 6, human papillomavirus and human variola virus as well as two species of rodent retroviruses and an insect polyhedrosis virus.

Discussion

The analysis of these subtraction libraries indicates that schizophrenia may be associated with the increased transcription of a range of RNAs in brain frontal cortex. The increases in RNA transcription did

Table 3. Sequences of unknown function (expressed sequence tags)

Genbank ascension No.	No. matches	Chromosomal localization*
R13524	3	11
T33605	2	6
H45647	2	8
W95882	2	21
T06692	1	1
AA127250	1	1
AA533286	1	2
AA122420	1	4
W74189	1	7
AA653775	1	11
AA135797	1	11
R37789	1	13
W72666	1	21
N30306	1	X
AA488130	1	X
AA441834	1	
AA22902	1	
R13857	1	
AA287675	1	
AA045204	1	
H20646	1	
AA169656	1	
AA046458	1	
AA322788	1	
AA322648	1	
AA329697	1	
AA514960	1	
z44058	1	
AA465704	1	
R78606	1	
N93621	1	
AA074356	1	
R14693	1	

*Chromosomal localization as indicated by the Unigene Program, National Library of Medicine, National Institutes of Health, Bethesda, Md. Blank space indicates that the chromosomal localization has not as yet been determined

not appear to be random but rather were concentrated on markers of cellular activation and immunity. For example, we found increases in RNAs encoding proteins with homology to mitochondrial and heat shock proteins. These findings are consistent with previous studies of RNA expression in the brains of individuals with schizophrenia [23,24]. We also found evidence of upregulation of RNAs capable of encoding proteins involved a number of cellular metabolic and immune-mediated pathways. The potential role of these RNAs and proteins in the pathogenesis of schizophrenia is the subject of ongoing investigations.

Table 4. Novel Sequences

Genbank ID	No. matches	Viral homology (predicted protein)
AF017090	2	Autographa californica nuclear polyhedrosis virus
NA	1	Retrovirus-rat
NA	1	Retrovirus-mouse
AF017094	1	Human Variola (Smallpox) virus
NA	1	Lactic Dehydrogenase Virus
NA	1	Human Herpes Virus type 6
AF017092	1	Human papillomavirus type 25
AF017086	1	
AF017087	1	
AF017089	1	
AF017091	1	
AF017093	1	
AF017095	1	
NA	1	
NA	1	
NA	1	
NA	1	
NA	1	

NA Genbank Ascension Number not yet assigned. Sequence is available from the authors

We also identified a number of sequences which did not match known human proteins but which contain open reading frames predicted to encode proteins homologous to those of pathogenic human and animal viruses. These RNAs may represent genomic or messenger RNA derived from novel viral agents. Of interest in this regard is that fact that several of the predicted proteins are homologous to viruses of non-human origin. This finding may represent the infection of humans with xenotropic viruses or with previously unrecognized human viruses with homology to viruses of non-human origin. Of particular interest in this regard would be the identification of human retroviruses with homology to animal retroviruses since such agents have recently been proposed to play an etiological role in other chromic human disease such as multiple sclerosis [25] and diabetes [26]. Alternately, it is possible that these RNAs encode previously uncharacterized human proteins with sequence homologies to viral polypeptides. The possible role of these viral or cellular proteins in the etiology of schizophrenia is the subject of ongoing investigations.

Our studies indicate that schizophrenia is associated with the upregulation of a number of different RNA species. These RNAs are thus unlikely to represent single genetic determinants of schizophrenia but rather may reflect broad alterations in cellular function within the central nervous system. It is also likely that the altered levels of RNA transcription are modulated by multiple genetic and environmental factors. These findings are thus consistent with the identification of the non-random sharing of genetic determinants in individuals with schizophrenia but the corresponding inability to definitively associate individual genes with disease phenotype using classical genetic linkage methodologies. The development of methods for the integration of both genetic and environmental determinants might lead to increased precision in terms of the identification of genes associated with disease susceptibility.

Our studies indicate that subtraction hybridization methodologies using post-mortem brain tissue obtained from individuals with schizophrenia can identify RNAs with potential etiological significance. Ongoing studies include the further localization and functional characterization of the RNAs found in the library as well as the documentation of the differential expression of the RNAs by independent methods such as reverse transcriptase polymerase chain reaction and Northern blotting. In addition, we are analyzing other brain areas such as the hippocampus to determine if the same RNAs are differentially expressed in additional anatomic sites. We are also performing studies with subtraction performed in the reverse orientation in order to identify RNAs that are down-regulated in individuals with schizophrenia. Finally, we are performing similar analyses of subtraction libraries derived from the brains of individuals with bipolar disorder. The analysis of brain tissues by subtraction and other genomic methodologies may lead to the identification of both genetic and environmental determinants of RNA transcription and hence to the development of new methods for the understanding, diagnosis, and eventual treatment of schizophrenia and other serious human neuropsychiatric diseases.

References

1. Barondes SH, Alberts BM, Andreasen NC, Bargmann C, Benes F, Goldman-Rakic P, Gottesman I, Heinemann SF, Jones EG, Kirschner M, Lewis D, Raff M, Roses A, Rubinstein J, Snyder S, Watson SJ, Weinberger DR, Yolken RH (1997) Workshop on schizophrenia [Meeting Report]. Proc Natl Acad Sci USA 94(5):1612–1614
2. Portin P, Alanen YO (1997) A critical review of genetic studies of schizophrenia. II. Molecular genetic studies [Review]. Acta Psychiatr Scand 95(2):73–80
3. Tsuang MT, Faraone SV (1995) The case for heterogeneity in the etiology of schizophrenia [Review]. Schizophr Res 17(2):161–175
4. Moldin SO, Gottesman II (1997) At issue-genes, experience, and chance in schizophrenia-positioning for the 21st century. Schizophr Bull 23(4):547–561
5. Yolken RH, Torrey EF (1995) Viruses, schizophrenia and bipolar disorder. Clin Microbiol Rev 8:131–145

6. Torrey EF, Miller J, Rawlings R, Yolken RH (1997) Seasonality of births in schizophrenia and bipolar disorder: a review of the literature. Schizophr Res 28:1–38
7. Thursz MR (1997) Host genetic factors influencing the outcome of hepatitis. J Viral Hepatitis 4(4):215–220
8. Li X, Quan Y, Wainberg MA (1997) Controlling elements in replication of the human immunodeficiency virus type 1. Cell Mol Biol 43(3):443–454
9. Yolken RH, Torrey EF (1995) Viruses, schizophrenia and bipolar disorder. Clin Microbiol Rev 8:131–145
10. Yee F, Yolken RH (1997) Identification of differentially expressed RNA transcripts in neuropsychiatric disorders. Biol Psychiatry 41(7):759–761
11. Johnston NL, Cerevnak J, Shore AD, Torrey EF, Yolken RH (1997) Multivariate analysis of RNA levels from postmortem human brains as measured by three different methods of RT-PCR. J Neurosci Methods 77(1): 83–92
12. Ukyanov K, Diatchenko L, Chenchik A, Nanisetti A, Siebert P, Usman N, Matz M, Lukanov S (1997) Construction of cDNA libraries from small amounts of total RNA using the suppression PCR effect. Biochem Biophys Res Commun 230:285–188
13. Diatchenko L, Lau Y-FC, Campbell AP, Chenchik A, Moqadam, F, Huang B, Lukyanov S, Lukyanov K, Gurskaya N, Sverdlov ED, Siebert PD (1996) Suppression subtractive hybridization: a method for generating differentially regulated or tissue-specific cDNA probes and libraries. Proc Natl Acad Sci USA 93:6025–6030
14. Adams MD, Kelley JM, Gocayne JD, Dubnick M, Polymeropoulos MH, Xiao H, Merril CR, Wu A, Olde B, Moreno RF (1991) Complementary DNA sequencing: expressed sequence tags and human genome project. Science 252(5013):1651–1656
15. Boguski MS, Lowe TM, Tolstoshev CM (1993) dbEST–database for "expressed sequence tags". Nature Genet 4(4):332–333
16. Sedgwick JD (1995) Immune surveillance and autoantigen recognition in the central nervous system. Aust NZ J Med 25(6):784–792
17. Huntriss JD, Twomey BM, Isenberg DA, Latchman DS (1994) Enhanced transcription of the gene encoding the SmN autoantigen in patients with systemic lupus erthematosus does not result in enhanced levels of the SmN protein. Autoimmunity 19(2):81–87
18. Straub RE, MacLean CJ, Kendler KS (1996) The putative schizophrenia locus on chromosome 6p: a brief overview of the linkage studies. Mol Psychiatry 1(2):89–92
19. Couillin P, Le Guern E, Vignal A, Fizames C, Ravise N, Delportes D, Reguigne I, Rosier MF, Junien C, van Heyningen V, et al. (1994) Assignment of 112 microsatellite markers to 23 chromosome 11 subregions delineated by somatic hybrids: comparison with the genetic map. Genomics 21(2):379–387
20. Fananas L, Fuster C, Guillamat R, Miro R (1997) Chromosomal fragile site 1q21 in schizophrenic patients. Am J Psychiatry 154(5):716
21. Kendler KS, MacLean CJ, O'Neill FA, Burke J, Murphy B, Duke F, Shinkwin R, Easter SM, Webb BT, Zhang J, Walsh D, Straub RE (1996) Evidence for a schizophrenia vulnerability locus on chromosome 8p in the Irish Study of High-Density Schizophrenia Families. Am J Psychiatry 153(12):1534–1540

22. Dann J, DeLisi LE, Devoto M, Laval S, Nancarrow DJ, Shields G, Smith A, Loftus J, Peterson P, Vita A, Comazzi M, Invernizzi G, Levinson DF, Wildenauer D, Mowry BJ, Collier D, Powell J, Crowe RR, Andreason NC, Silverman JM, Mohs RC, Murray RM, Walters MK, Lennon DP, Crow TJ et al. (1997) A linkage study of schizophrenia to markers within Xp11 near the MAOB gene. Psychiatry Res 70(3):131–143
23. Goodman AB (1994) Elevated risks for amyotrophic lateral sclerosis and blood disorders in Ashkenazi schizophrenic pedigrees suggest new candidate genes in schizophrenia. Am J Med Genet 54(3):271–278
24. Whatley SA, Curti D, Marchbanks RM (1996) Mitochondrial involvement in schizophrenia and other functional psychoses. Neurochem Res 21(9): 995–1004
25. Perron H, Garson JA, Bedin F, Beseme F, Paranhos-Baccala G, Komurian-Pradel F, Mallet F, Tuke PW, Voisset C, Blond JL, Lalande B, Seigneurin JM, Mandrand B (1997) Molecular identification of a novel retrovirus repeatedly isolated from patients with multiple sclerosis. The Collaborative Research Group on Multiple Sclerosis. Proc Natl Acad Sci USA 94(4):7583–7588
26. Conrad B, Weissmahr RN, Boni J, Arcari R, Schupbach J, Mach B (1997) A human endogenous retroviral superantigen as candidate autoimmune gene in type 1 diabetes. Cell 90(2):303–313

Correspondence: Dr. R.H. Yolken, Stanley Division of Developmental Neurovirology, Johns Hopkins University School of Medicine, 600 N. Wolfe Street, Blalock 1111, Baltimore, MD 21287-4933, U.S.A.

Cerebrospinal fluid filtration in a case of schizophrenia related to "subclinical" Borna disease virus encephalitis

K. Bechter[1], **S. Herzog**[2], **V. Schreiner**[1], **K-H. Wollinsky**[3], and **R. Schüttler**[1]

[1] University of Ulm, Department Psychiatry II, and Department of Psychiatry, Bezirkskrankenhaus Günzburg,
[2] Institute of Virology, Justus-Liebig-University, Gießen, and
[3] Department of Anaesthesiology and Intense Care, Rehabilitation Hospital Ulm, University of Ulm, Federal Republic of Germany

Summary

Schizophrenia may represent a clinical syndrome caused by various etiologies and through various pathogenetic pathways. Recent studies suggest viruses to be a more important cause of schizophrenia than previously thought. Borna disease virus (BDV) causes meningoencephalitis or behaviour abnormalities in a broad range of animals, and accumulating data suggest a causal role of BDV for human neuropsychiatric disorders including schizophrenia. However, it is difficult to establish causality in the clinical situation. We addressed this problem by detailed investigations over the long term in a suggestive single case. Furthermore, we hypothesized a therapeutic effect of cerebrospinal fluid filtration (CSFF) in such case comparable to that in therapy resistant postinfectious Guillain-Barré syndrome (GBS).

A variety of signs and findings in a 28 years old man indicated a subclinical BDV encephalitis being responsible for DSM-IV recent onset schizophrenia, mainly: elevating BDV specific serum ab's, elevated CSF proteins, slow alpha-waves and dysrhythmia in EEG on admission (no gross neurologic abnormalities). Positive schizophrenic symptoms improved under psychopharmaca medication but not the primary negative symptoms. Therefore the patient was treated by CSFF in two filtration series, each over five days. Outcome was monitored by a battery of assessments and tests.

Negative schizophrenic symptoms were rapidly improved under CSFF, in parallel neurological soft signs, performance in attention tests and EEG; this was similar in time pattern as GBS symptoms improve under CSFF. Clinical status and test performance remained on an improved level after CSFF ($p < 0.05$) and improved further spontaneously over the long-term (8 months inpatient, 11 months outpatient observation) to full remission.

We conclude, that we probably observed the natural course of a mild or 'subclinical' BDV-encephalitis-related schizophrenia, which could be ameliorated by psychopharmaca and CSFF. Removing yet partially defined toxic factors, similar to that found in GBS, from cerebrospinal fluid may explain clinical improvement under CSFF. 'Subclinical' encephalitis by BDV or other viruses may more frequently underly schizophrenic symptoms than previously detected. Our observation indicates a novel pathogenetic principle in a case of schizophrenia.

I. Introduction

Borna disease virus is a frequent cause of meningoencephalitis in horses and sheep in middle Europe, caused by the neurotropic negative stranded RNA virus BDV (de la Torre et al., 1994; Schneider et al., 1994). According to recent studies, BDV is widely distributed over the world in a broad host range, although Borna disease (BD) was earlier described in Europe only (Rott and Becht, 1995). Pathogenesis of BDV encephalitis is understood as a T cell-dependent immune mechanism, the preferential site being the hippocampus. A mild form of BD in animals can result exclusively in behavioural disturbances. Therefore a causal relationship to human psychiatric disorders has been suggested (Rott et al., 1985; Amsterdam et al., 1985). Comparably large epidemiologic studies in humans showed an increased prevalence of BDV serum antibodies in psychiatric and possibly neurologic patients, especially in young psychiatric patients (Bechter et al., 1998a). However, the question of causality of BDV for human psychiatric disorders is yet open and there are considerable discrepancies in findings and interpretation (Bechter et al., 1996; Hatalski et al., 1997; Richt et al., 1997a,b).

Earlier we isolated BDV or a related agent from the cerebrospinal fluids (CSF's) of two neurologic patients with lymphocytic meningoencephalitis and of one psychiatric patient with acute schizophrenia (Rott et al., 1991). Recently BDV specific RNA has been reported in post mortem human brains from patients suffering from schizophrenia or affective psychosis or cognitive disturbances during lifetime (de la Torre et al., 1996; Salvatore et al., 1997; Haga et al., 1997). However, positive results from immunohistopathology are lacking and BD virus has not yet been isolated from the human brain. – There are controversial results from investigations of the peripheral blood: Some groups reported BDV specific RNA in up to 50% of psychiatric patients with depression or schizophrenia (Bode et al., 1996, Sauder et al., 1996), whereas other groups in Germany and USA were not able to detect BDV RNA in peripheral blood leukocytes in large samples of patients and suggested laboratory contamination as 'false' positive reactions (Richt et al., 1997b; Herzog et al., 1997; Lieb et al., 1997). For the indirect immunoflurescence assay (IIFA) as used in our human studies, high specificity and sensitivity was demonstrated in parallel studies on horses (including brains, sera and CSF's) naturally infected with BDV (Herzog et al., 1994). It is important to note that only in horses, not in laboratory animals, BDV serum antibody (ab) titers (determined by IIFA) are as low as in humans. In ongoing human studies over years we found in 25% of selected BDV seropositive psychiatric inpatients (n > 40), most of them suffering from acute schizophrenia or affective psychosis elevated BDV specific immunoglobulin G (IgG) in CSF, indicating an active although slight BDV encephalitis (Bechter et al., 1995). Now we observed a single case in whom a slight BDV encephalitis was probably the cause of recent onset schizophrenia with lasting negative symptoms. Because meningoencephalitis or postencephalitic Guillain-Barré syndrome (GBS) recently can be treated by CSF filtration (CSFF; Wollinsky et al., 1994), we introduced this technique in this single case of schizophrenia (Bechter et al., 1998b). Our observations shed new light on the possible causality and pathogenesis of schizophrenic symptoms, at least in a subgroup of cases.

II. Methods

Patient

A previously healthy 28 year old man was admitted for the first time for psychiatric inpatient treatment because of a gradual change of personality since about 3 months with severe suicidal ideation. In the patient's subjective experience initial symptoms were loss of energy, drive and performance. On admission he showed slow formal thought, psychomotor slowing, reduced concentration and reduced spontaneity. Furthermore he showed a mixed depressive-agitated mood, inadequate affect and bizarre delusional ideas. Day-night-rhythm was severely disturbed. No gross neurologic abnormalities were found, just the following neurological soft signs: Slow motion of the arms when walking, slow somewhat straddling gait, bradydiadochokinesis and hypomimia. He complained of head pressure, but no headache and he did not show meningism. Body temperature was normal as was the further clinical examination. Routine laboratory investigations (blood, urine) showed no significant abnormalities. Serology for the full range of neurotropic infectious agents was normal, except positive BDV serum IgG ab's with a titer of 1:20 short after admission. **CCT** (Siemens Somatom plus) and **MRI** (Siemens-1.0 Tesla Magnetom Expert) of the brain, including sagittal, coronal and transversal T_1 and T_2 weighed sequences, a flair sequence and a Gadolinium contrast sequence, showed normal findings. A diagnosis of recent onset schizophrenia was made (DSM IV 295.30; ICD-10F 20.09). The patient responded to neuroleptic/antidepressant medication by rapid normalization of disturbed day-night rhythm, positive symptoms improved moderately, negative symptoms did not respond. Medication: flupentixol up to 10 mg, amitriptyline up to 100 mg, lorazepam up to 5 mg (orally, per day). When BDV encephalitis was suggested,

200 mg carbamazepine was added. Medication remained unchanged from 3 weeks before first CSFF series.

Documentation

The patient was treated as an inpatient for 8 months in our psychiatric department and as an outpatient after. Standardized diagnostic instruments on psychopathology applied included BPRS, SCL-90-R, PANSS, PD-S, HAMD and MADRS. IQ was tested by HAWIE-R and by SPM. A psychiatric interview including specifically neurological soft signs examination was documented on video tape repeatedly. Information processing speed was measured by the following tests: Aufmerksamkeits-Belastungstest (d2-test), Zahlenverbindungstest (ZVT), Stroop Color-Word-Interference Test. Standardized tables from German population were available for d2 (a letter cancellation test) and ZVT (a number connection test). Test performances during four time segments were analyzed statistically (before and after 1st fs, after 2nd fs, and at discharge; a minimum of 4 measurements within each segment) by analysis of variance and Scheffé-test.

EEG

12 channel routine paper EEG's were performed repeatedly (n = 28) and evaluated by 3 raters (blind, independently). The spectra of EEG's taken before, during and after 2nd fs were analyzed by Neurofile Program, Nihon Kohden Europe, from artefact free EEG periods with a minimum of 60s duration eyes-closed-condition (mean 130.8 ± 60.0, range 60.0 – 170.9), calculated against common average (FP_1 + FP_2 excluded).

BDV immunoglobuline G (IgG) antibodies in serum and CSF were assayed with IIFA and immunoblot as described earlier (Herzog et al., 1994).

CSF filtration (CSFF) was performed as described earlier (Wollinsky et al., 1994). In short: By a lumbar puncture a 16-G-catheter was introduced into the lumbar subarachnoid space under sterile conditions and attached to the Pall-CSF-Filter system (Pall GmbH Biomedizin, Dreieich). Automatized suction and reinfusion of CSF was done by a bidirectional computerized syringe pump, flow- and pressure-controlled (Infors, Bottmingen, Switzerland). Two filtration series (fs') each consisting of a daily filtration run (fr) in the morning for five successive days, were performed in a time distance of 3 weeks (see Fig. 1). A volume of 200–250 ml CSF was filtered in daily fr's, within 5–6 cycles at a rate of 3 ml CSF/min.

III. Proposing cerebrospinal fluid filtration (CSFF) as therapy

We report on the first CSFF performed in a psychiatric patient, therefore the reasons to propose such are discussed in detail:

a) Assessment of a BDV specific 'subclinical' encephalitis as the probable cause of schizophrenic psychosis: During the first 5 weeks of inpatient treatment we observed an increase of BDV specific IgG serum

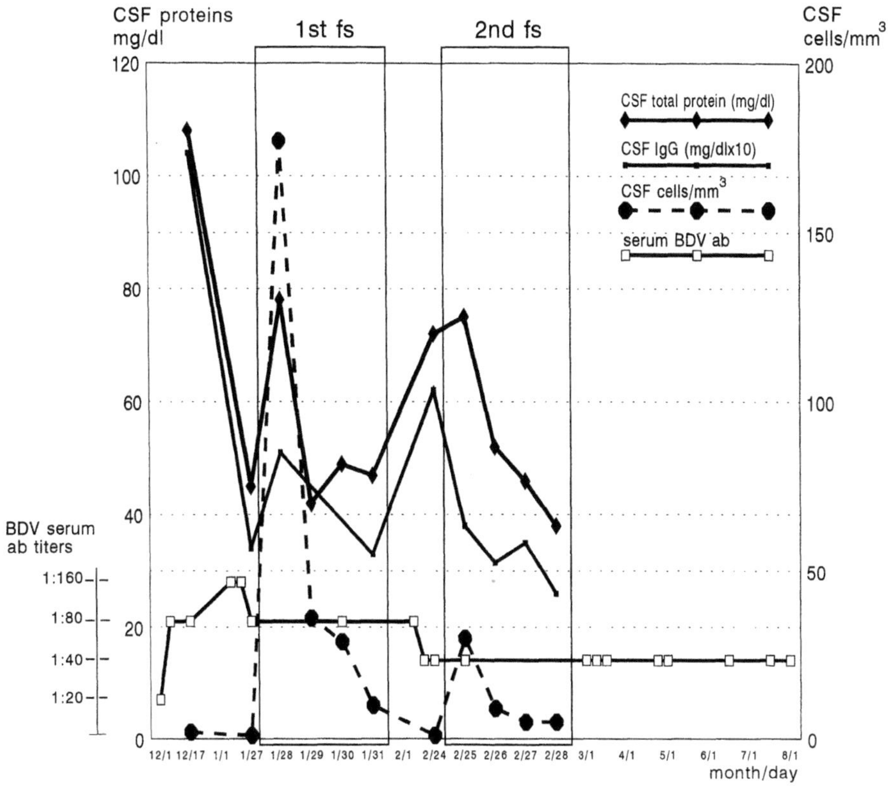

Fig. 1. CSF findings before and during both CSF filtration series (1st and 2nd fs). Represented are the morning values before daily filtration run. serum protein values were stable (not demonstrated). Development of BDV serum IgG antibody titers (BDV ab) is given in addition. Please note the graphically extended time frame during both CSF-fs'

ab's by 4 titer grades (from 1:20 to 1:160); specificity of IIFA results was confirmed by the immunoblot, sera recognizing the p-38 BDV antigen; initial EEG showed a slow 8–8.5 /s alpha rhythm, interposed were frequent theta waves and some dysrhythmic groups; CSF proteins were elevated about 3-fold above normal when investigated in the 2nd week of inpatient treatment (see Fig. 1), demonstrating a proportional blood-CSF-barrier-disturbance according to the Reiber graph (Reiber and Felgenhauer, 1987); BDV specific IgG was shown in CSF after 5-fold concentration. CCT and MRI of the brain were normal; but this is a frequent finding in clinical meningitis (Stark and Bradley, 1992) and compatible with the microscopic features of BD lesions and the limits of MR sensitivity (Bechter et al., 1994), and therefore did not contradict against our diagnostic conclusion.

b) CSFF can immediately improve paresis in acute (or chronic) GBS and is used in ongoing controlled studies accepted by the ethics committee of the University of Ulm for treating GBS or single cases of menginoencephalitis (Wollinsky et al., 1994). Most GBS cases likely represent a postinfectious autoimmune disorder. The pathogenesis of BDV encephalitis however has been characterized as a delayed type immunopathologic reaction (Bilzer and Stitz, 1996), and a spontaneous autoimmune shift has been observed in animals (Rott et al., 1993). We therefore hypothesized a pathogenetic similarity to GBS and a similar therapeutic effect in this probably BDV encephalitis related schizophrenia. A prerequisite was, that by lumbar CSFF the CSF of subarachnoid spaces around the brain and of the ventricles can be filtrated (Pfausler et al., 1995).

c) Our patient suffered strongly from suicidal ideation related to the self-experienced primary negative symptoms, from which he desired urgently relief. For primary negative schizophrenic symptoms no proven treatment exists generally (American Psychiatric Association, 1997), and these symptoms were not improved by previous treatment. (The risk of trying activating antidepressants possibly aggravating suicidal ideation appeared at that point of time not balanced with a small possible effect on the clinical status.) Furthermore, the effect of CSFF in neurologic patients, can be considered somehow "curative" and such was hoped in this recent onset schizophrenia.

Safety

About 3–4000 CSFF's in more than 400 neurologic patients have now been performed since introduction of the method and no severe side effects have been observed yet. The risk of bacterial contamination of the subarachnoid spaces is controlled by sterile conditions, in the case of contamination treatment by antibiotics was possible. Besides, CSFF is used as an adjuvant treatment in bacterial meningoencephalitis (Pfausler et al., 1995; Brizzi et al., 1996; Kalischweski et al., 1997). Other risks were just that of lumbar puncture, a frequent diagnostic procedure.

Informed consent

Premorbid personality was of a high intellectual level (SPM-IQ 127, HAWIE-R-IQ 132, when measured at discharge) and the patient was fully able to understand our explanations. He gave written informed consent being aware of the new approach, in accordance with the regulations for an individual treatment trial. The patient was able to cooperate with the technical needs as with the manifold assessments and tests. He accepted transmission to the neurologic intense care ward for one week during each filtration cycle. Informed consent was further obtained from his relatives.

IV. Results

Clinical course during and after CSFF

With day 3 of the 1st fs an impressive improvement of clinical symptomatology was observed: Psychomotor performance was considerably quicker, eg. diadochokinesis accelerated by 20%, reaching a level similar to that at discharge according (time measurements on the video records). He was more alert and active. He reported improvement from head pressure and from feeling heavy. Test performance improved significantly and remained on a higher level from this point of time ($p < 0.05$; see Fig. 2). Relatives reported that from day 3 of 1st fs he was spontaneously talking to them, not doing so in the 5 months before. Because of the good effect a 2nd fs was performed 3 weeks

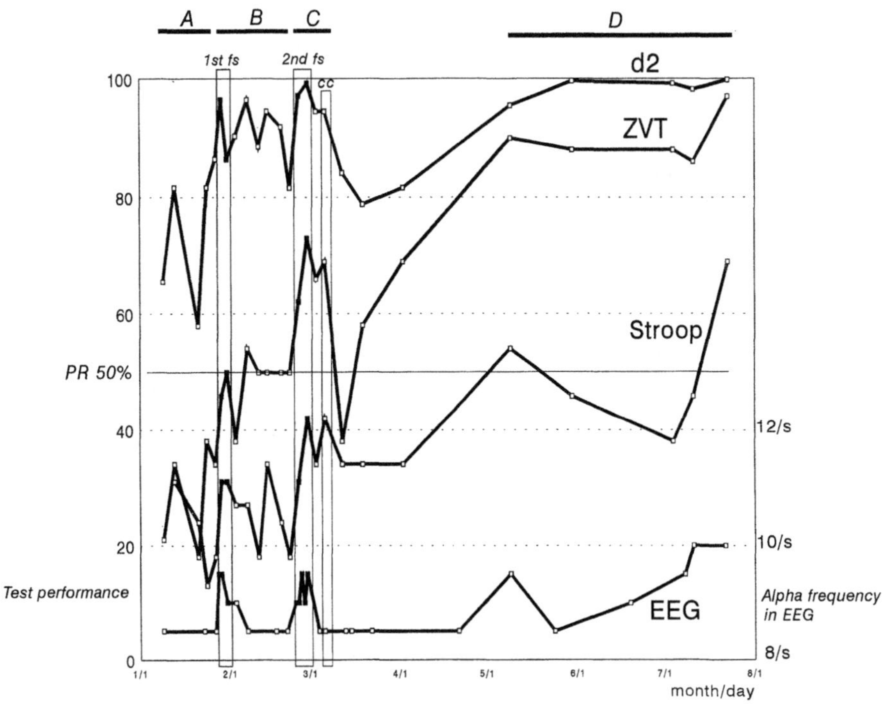

Fig. 2. Development of patient's test performance (d2-Test, ZVT, Stroop-interference-task) and EEG basic rhythm (α-frequency) including time before, during and after CSF filtration series. Test performance is given in percent ranks (PR) of age-corresponding normal population. Frames delineate time segments during the following therapeutic procedures: *1st fs* first filtration series. *2nd fs* second filtration series, *cc* corticoid-treatment. Tests performed during fs' are marked (closed squares). Upper lines delineate time segments statistically evaluated (see Results): *A* before 1st fs, *B* during and after 1st fs, *C* during and after 2nd fs, *D* at discharge

later. With the 2nd fs the patient's feelings of head pressure disappeared completely, social avoidance behavior and test performance ($p < 0.05$) improved further (see Table and Fig. 2). After 2nd fs a high dose of corticoides was given for 4 days (500 mg methylprednisolone a day, intravenously) to possibly stabilize the clinical status. However, patient rather deterioriated for a time period of about one week (see Fig. 2). – In the next 2 months the patient improved spontaneously slightly further; psychotropic medication was then reduced a little. Although the patient was not fully remitted yet and was unable to take over his earlier occupation, he nevertheless wanted to be discharged. After discharge he did not take any medication further, deterioriated and came back to inpatient treatment 18 days later: day-night-rhythm was again disturbed (moderately), positive and negative schizophrenic symptoms slightly more intense (see also Fig. 2). Now the following neuroleptic/antidepressant medication (fluphenazine 3 mg, amitriptyline 75 mg, chlorprothixen 50 mg orally, daily) was given. The patient responded again on medication, day-night-rhythm normalized within 2 days, and he showed a status similar to that 20 days before. He then joined an inpatient rehabilitation program for the next months. For 6 weeks he was additionally prescribed 100 mg fluvoxamine but no positive effect observed and fluvoxamine was stopped. Nevertheless, the patient improved gradually further, most obviously during the last 6 weeks of inpatient treatment. At discharge he was in a fully remitted status (see Fig. 2). However, in a tentative 5 days off-medication phase shortly before discharge the patient deterioriated again slightly (negative symptoms, sleep disturbance), not the EEG. Therefore previous medication was again prescribed. After discharge the patient took over his earlier occupation successfully. An outpatient catamnesis of now 11 months is very good (confirmed by patients' relatives), patient being fully working and receiving no medication since 9 months.

EEG

At the beginning of inpatient treatment EEG showed a 8–8.5 /s irregular alpha-rhythm, frequently interposed theta-waves and dysrhythmias forced during and after hyperventilation. Medication was of no obvious influence on the EEG in 3 off-medication phases: early (12/12–13/12), in the midst (4/4–21/4) and late (5/7–11/7) during the 8 months of inpatient treatment. EEG basic rhythm apparently showed a spontaneous change over the observed time period: from initial slow somewhat irregular alpha-basic rhythm to an (accelerated) regular alpha-rhythm at discharge (Fig. 2), initial frequent theta waves and dysrhythmia declining and eventually disappearing in parallel to alpha improvement (not demonstrated). Similar to this long-term changes, during both CSFF series the EEG improved even after single fr's: alpha

Table 1. Patient' sum scores in various standardized psychiatric rating scales tested before and after 1st CSF filtration series, after 2nd filtration series and at discharge

Scales	before CSF filtration	after 1st filtration series	after 2nd filtration series	at discharge
BPRS	**3.8**	**3.5**	**3.2**	**2.1**
Brief Psychiatric Rating Scale				
– ANDP: Anxiety/Depression	4.5	4	3.5	2.3
– ANER: Anergia	4.8	4.5	3.8	2
– THOT: Thought Disturbance	4	3.8	3.8	2
– ACTV: Activation	4	3.3	3.3	2.7
– HOST: Hostile-Suspiciousness	1.3	1.3	1.3	1.3
SCL-90-R	**1.3**	**0.6**	**0.4**	**0.1**
Self-Report Symptom Inventory				
– somatization	1.5	0.3	0.1	0.1
– obsessive-compulsive	1.3	0.5	0.2	0
– interpersonal sensitivity	0.8	0.2	0.2	0.1
– depression	2.5	1.5	1.2	0.4
– anxiety	1.2	0.5	0.4	0.1
– anger-hostility	0.5	0	0	0
– phobic anxiety	0.1	0	0	0
– paranoid ideation	0.5	0.5	0.3	0
– psychoticism	1.1	0.6	0.4	0
– additional items	2.1	1.4	1.1	0.3
PANSS				
Positive and Negative Syndrome Scale				
– Positive Scale	18	17	17	10
– Negative Scale	27	23	21	12
– General Psychopathology Scale	41	36	35	22
PD-S				
Paranoid-Depressiveness-Scale				
– P-Scale	7	3	3	0
– D-Scale	24	17	15	1
HAMD	**28**	**24**	**21**	**5**
Hamilton Depression Scale				
MADRS	**40**	**24**	**19**	**5**
Mantgomery-Asberg-Depression-Rating Scale				

For BPS and SCL-90-R sum-score / number of items, for the other scales the sum scores are given. Scores (or quotients') ranges (minimal to maximal): BPRS 1–7; SCL-90-R 0–4; PANSS Positive Scale 7–49, Negative Scale 7–49 and General Psychopathology Scale 16–112; PD-S 0–48; HAMD 0–67; MADRS 0–60

Table 2. Representation of significant improvements (acc. analysis of variance and Scheffé test) of test performance in percent ranks during 8 months inpatient treatment of the patient comparing 4 time segments A,B,C,D before and after CSF filtration cycles.: + = $p < 0.05$, 0 = not significant. Insignificant values in C/D, D/B and D/C of d2-test were likely explanined by reaching upper limit of performance in B/A and C/A, (see Fig. 2 and calculation of absolute values (data not shown)

Time segments compared to segments	B	C		D		
	/A	/A	/B	/A	/B	/C
d2-Test [F(17.3) = 13.67; sig of F = 0.0001]	+	+	0	+	0	0
ZVT [F(18.3) = 101.19; sig of F = <0.00001]	+	+	+	+	+	+
Stroop [F(18.3) = 13.09; sig of F = 0.0001]	0	+	0	+	+	0

Time segments: *A* before first filtration series, *B* during and after first filtration series, *C* during and after second filtration series, *D* time at discharge (compare Fig. 2)

frequency was accelerated by 0.5–1.5 /s (Fig. 1), power increased and theta waves were fewer (also shown in a spectral analysis of the data; not demonstrated). EEG improvement after single fr's was not lasting (during both f's), that means EEG was again deterioriated the next morning after single fr's. However inprovement became more prominent with increasing number of fr's during both weeks.

Cerebrospinal fluid:

CSF proteins (albumin, IgG, total protein) were initially (date 17/12) elevated about 3-fold above normal (Fig. 1; see also II; these measures were independently confirmed in the laboratories of the clinics of Günzburg and of Augsburg). Cell count, glucose, and colour were normal, no oligoclonal bands found. – When beginning CSFF (10th week of inpatient treatment) surprisingly the CSF proteins had nearly normalized. During 1st fs a strong increase of cells and proteins (and 2nd fs), lowering at the end of the fs' was observed; this was expected from experiences with CSFF in GBS (Wollinsky et al., 1994). Increase of cells was only weak during the 2nd fs.

Tests

Test performance improved during and after both fs' to high levels (>2 standard deviations in comparable population), the levels reached remained widely stable thereafter ($p < 0.05$; Fig. 2); reaction time improved from 206 ± 36.8 ms to 185 ± 22.2 during 2nd fs (1st fs not tested). Values above the mean were reached in all tests at the end of inpatient treatment.

Adverse effects of CSF filtration

No severe adverse effects were observed. When beginning the daily fr, a feeling of cold in the back for several minutes was reported. The patient further complained about the sober atmosphere of the intensive care ward and therefore he was allowed to stay on the psychiatric ward for some hours daily, the lumbar catheter being left on site closed under sterile dressing.

V. Discussion

The etiopathogenesis of affective and schizophrenic psychoses remains unknown today. Viral infections might explain more cases of psychosis than previously thought: With improved methods of CSF investigation some cases were attributed to probable slight viral encephalitis by known viruses (Torrey, 1986); a recent study on identical twins showed evidence that a considerable part of schizophrenic and affective psychoses could be related to viral causes generally (Torrey et al., 1994); in about 40% of schizophrenic patients a decreased CSF flow has been found and a relationship to unknown subclinical viral infections was suggested (Oxenstierna et al., 1996); in about 25% of patients with schizophrenic or affective psychosis a slight increase of CSF proteins (Müller and Ackenheil, 1995) or disturbances in cytokine levels (Maes et al., 1994) have been reported, compatible with an autoimmune or a viral etiology.

One possible viral cause of psychiatric disorders recently discussed is BDV, however the possible pathogenic impact remained elusive (Bechter et al., 1996; Hatalski et al., 1997; Richt et al., 1997a,b). Rare "lucky" cases played a key role to suggest a causal relationship between syphilis and general paresis (Bechter, 1995). We think, that the findings and observations in our case could play a key role for a better understanding of cases of schizophrenia possibly caused by subclinical viral encephalitis in general and to elucidate a possible pathogenic role of BDV for psychiatric disorders specifically: The first question in our case was, whether the onset of schizophrenic psychosis was related to BDV encephalitis. We concluded that probably yes (see III). However, the patient was seen as an inpatient only 3 months after gradual onset of symptoms. So the observed increase of BDV specific IgG in serum during the first weeks of inpatient treatment might have been unrelated to the onset of disease; on the other hand, clinical symptoms were indeed deterioriating during this time period and objective measures and findings indicated a subclinical BDV encephalitis in parallel: CSF proteins were elevated, EEG was slow and dysrhythmic, patient showed neurological soft signs. So the time period when inpatient treatment begun can be understood as the culmination of a protracted subclinical BDV encephalitis. Furthermore, all these symptoms and findings reversed (improved) during the 8 months inpatient treatment in parallel or in a time sequence pattern compatible with our interpretation. And similar time courses have been observed in natural BD in animals (Rott and Becht, 1995). Summing up, we probably observed the natural course of a human BDV encephalitis causing a psychotic disorder classified as schizophrenia, and delayed recovery from. It is possible, that an increased liability to become schizophrenic preexisted in the patient and it remains open whether BDV encephalitis was reactivated or newly acquired.

Secondly, we observed an unknown therapeutic effect of CSFF on negative schizophrenic symptoms, in parallel on neurological soft signs, on clinical signs such as head pressure, on EEG disturbances and on test performance. The time pattern of improvement was similar to that in GBS or MS cases under CSFF (Wollinsky et al., 1994; Haas et al., 1995). Or, could a placebo effect explain the improvement during CSFF? This is unplausible from several reasons: the long-lasting severe primary negative schizophrenic symptoms were not improved by medication or psychological treatments before; the degree of alpha-acceleration of EEG (and in parallel a decline of interposed theta waves) even after single fr's, the extent of improvement in test performance, and the improvement of neurological soft signs in parallel can hardly be explained by placebo in itself. When further considering that these concerted improvements of signs and findings during CSFF were differentiated over time (the more simple d2 and ZVT-tests improved even in 1st fs, the more complicated Stroop in the 2nd fs), and that a certain stability on higher performance levels was reached after both fs' ($p < 0.05$), a causal interdependence of these concerted short-term changes embedded in a similar long-term improvement, is strongly suggested and cannot be explained by placebo. Learning effects are unlikely to explain improvement in test performance, because such are reported low in the tests applied and because of the large time gaps in between testing. Can CSFF-improved negative symptoms be explained by a wash-out of medication? Such can be excluded because (negative and other) symptoms were deterioriated during three off-medication phases (see results). In conclusion: a "subclinical" protracted BDV encephalitis with lasting negative symptoms could well fully explain the symptoms of this recent onset schizophrenia as the course of disease and make understandable the positive effects of CSFF.

How did CSFF work in our case? Cells and a variety of soluble factors can be removed from CSF with high efficiency by CSFF (Rother et al., 1995). In demyelinating disorders a soluble toxin acting as a sodium channel blocker has been found (Brinkmeier et al., 1996). Activated immune cells seem to play a crucial role in GBS pathogenesis (Sivieri et al., 1997), and a prominent role of activated immune cells is established in pathogenesis of BD in animals (Bilzer and Stitz, 1996; Richt et al., 1997a). Because the increase of cells during 2nd fs was considerably smaller in our patient (Fig. 1) we speculate, whether the lasting positive effect of CSFF was related to removing activated immune cells resting, possibly producing toxic factors, within the CSF spaces around the brain (protein increases during CSFF are well known and may be understood as a reactive phenomenon on the injury from the catheter similar to observations after single lumbar punctures). Preliminary results (in preparation) from biochemical (high performance liquid chromatography = HPLC) and cell physiological analyses described earlier in Guillain-Barré syndrome (Brinkmeier et al., 1996) showed clear evidence of toxic factors being present in the CSF of our schizophrenic patient also. These findings strongly support our above interpretation of the case from the clinical data.

Slow alpha and increased theta activity is a frequent and most consistently reported EEG finding in acute schizophrenia (John et al., 1994). Cerebral flow studies suggest disturbed neuronal circuits underlying recent onset schizophrenia, possibly due to unknown toxic factors (Andreasen et al., 1997). In this context our observations, although made in a single case, might be of general interest. Borderline CSF protein elevations, similar to that found in our patient when beginning 1st fs (different to the strong CSF protein elevation found just 5 weeks before), were previously found in more BDV seropositive patients with acute psychosis (Bechter et al., 1995). So in this patient we may have been lucky to catch a time period with distinct CSF pathology or distinct CSF pathology may be present in few cases with a comparably strong immune reaction only (Bechter et al., 1996).

The observed positive effect of CSFF may be of interest also from another point of view: Recent prospective studies suggest a slight progressive brain atrophy taking place in the first years after onset of schizophrenia (de Lisi et al., 1997) or in a subgroup of schizophrenic patients (Nair et al., 1997). According to preliminary studies BDV seropositive psychiatric patients show more frequently (slight) brain atrophy than matched seronegative controls (Bechter et al., 1994; Waltrip et al., 1995), and BDV seropositive schizophrenic patients may represent a subgroup with more severe courses (Waltrip et al., 1997; Bechter et al., 1998a). Brain atrophy can occur after BD in animals (Bilzer and Stitz, 1996; Richt et al., 1997a). Regarding the unexpectedly good outcome in our patient and comparing the positive somehow "curative" effect of CSFF in GBS, meningoencephalitis or multiple sclerosis, we speculate, that CSFF could be a new therapeutic approach for a subgroup of schizophrenic patients, mainly with negative, possibly subclinical encephalitis related symptoms. It seems that unknown toxic factors can be removed from CSF (and from the brain) by CSFF. Improving sensitivity of methods to detect "subclinical" or, more correctly mild encephalitis appears to be a challenging research tool in schizophrenia research. The positive effects of CSFF in this case of schizophrenia were apparently due to removing toxic factors, which were similar to that found earlier in other chronic inflammatory CNS diseases (Brinkmeier et al., 1996; Köller et al., 1997), demonstrating a yet unknown pathogenetic principle in schizophrenia. CSFF may analogously provide a unknown curative treatment perspective at least in some cases of schizophrenia: Available neuroleptic treatments seem to be rather a "symptomatic" treatment because the overall course of schizophrenia seems to be unimproved by neuroleptics (van Os et al., 1997). However, the outcome of GBS can be improved by CSFF (Wollinsky et al., 1994; Köller et al., 1997), and we similarly suggest a possible improved outcome by CSFF treatment in "subclinical" encephalitis related schizophrenia.

Acknowledgements

We thank Prof. B. Widder and Dr. Breitinger, Department Neurology, BKH Günzburg; Dr. H. Durst, Department Psychiatry II, University of Ulm, BKH Günzburg; Dr. W. Behr, Central Clinics, Augsburg; for their help; and PD Dr. H. Brinkmeier and his coworkers P. Aulkemeyer and F. Weber for investigation of the patient's CSF by neurophysiological and biochemical methods.

References

American Psychiatric Association (1997) Practice guideline for the treatment of patients with schizophrenia. Am J Psychiatry 154[Suppl]4:42–43

Amsterdam JD, Winokur A, Dyson W, Herzog S, Gonzales F, Rott R, Koprowski H (1985) Borna disease virus: a possible etiologic factor in human affective disorders? Arch Gen Psychiatry 42:1093–1096

Andreasen NC, O'Leary DS, Flaum M, Nopoulos P, Watkins GL, Boles Ponto LL, Hichwa RD (1997) Hypofrontality in schizophrenia: distributed dysfunctional circuits in neuroleptic-naive patients. Lancet 349:1730–1734

Bechter K (1995) Research strategies in "slow" infections in psychiatry. Hist Psychiatry 6:503–511

Bechter K, Bauer M, Estler HC, Herzog S, Schüttler R, Rott R (1994) MRI findings in Borna disease virus seropositive psychiatric patients and controls: an extended study. Nervenarzt 65:169–174

Bechter K, Herzog S, Behr W, Schüttler R (1995) Investigations of cerebrospinal fluid in Borna disease virus seropositive psychiatric patients. Eur Psychiatry 10:250–258

Bechter K, Herzog S, Schüttler R (1996) Borna Disease Virus – possible cause of human neuropsychiatric disorders. Neurol Psychiatry Brain Res 4:45–52

Bechter K, Herzog S, Estler HC, Schüttler R (1998a) Increased psychiatric comorbidity in Borna disease virus seropositive psychiatric patients. Acta Psychiat Belg 98:190–204

Bechter K, Herzog S, Schreiner V, Wollinsky KH, Schüttler R (1998b) Negative symptoms improved by CSF filtration in a case of recent onset schizophrenia. Schizophr Res 29:157

Bilzer TH, Stitz L (1996) Immunopathogenesis of virus diseases affecting the central nervous system. Crit Rev Immunol 16:145–222

Bode L, Dürrwald R, Rantam FA, Ferszt R, Ludwig H (1996) First isolates of human Borna disease virus from patients with mood disorders. Mol Psychiatry 1:200–212

Brinkmeier H, Seewald MJ, Wollinsky KH, Rüdel R (1996) On the nature of endogenous antiexcitatory factors in the cerebrospinal fluid of patients with demyelinating disease. Muscle Nerve 19:54–62

Brizzi M, Thorén A, Hindfelt B (1996) Cerebrospinal fluid filtration in a case of severe pneumococcal meningitis. Scand J Infect Dis 28:455–458

de la Torre JC (1994) Molecular biology of Borna disease virus: prototype of a new group of animal viruses. J Virol 68:7669–7675

de la Torre JC, Gonzalez-Dunia D, Cubbitt B, Mallory M, Müller-Lantzsch N, Grässer FA, Hansen LA, Masliah E (1996) Detection of Borna Disease virus antigen and RNA in human autopsy brain samples from neuropsychiatric patients. Virology 223:272–282

de Lisi LE, Sakuma M, Tew W, Kushner M, Hoff AL, Grimson R (1997) Schizophrenia as a chronic active brain process: a study of progressive brain structural change subsequent to the onset of schizophrenia. Psychiatry Res Neuroimaging 74:129–140

Haas J, Sailer M, Düzel E, Tendolkar I, Wurster U (1995) CSF filtration: an experimental therapeutic approach in multiple sclerosis treatment. Neuropsychiatrie 9:109–111

Haga S, Motoi Y, Ikeda K and the Japan Bornavirus Study Group (1997) Borna disease virus and neuropsychiatric disorders. Lancet 350:592–593

Hatalski CG, Lewis AJ, Lipkin WI (1997) Borna disease. Em Infect Dis 3:129–135

Herzog S, Frese K, Richt JA, Rott R (1994) Ein Beitrag zur Epizootiologie der Bornaschen Krankheit beim Pferd. Wien Tierärztl Mschr 81:374–379

Herzog S, Pfeuffer I, Haberzettl K, Feldmann H, Frese K, Bechter K, Richt JA (1997) Molecular characterization of Borna disease virus from naturally infected animals and possible links to human disorders. Arch Virol [Suppl] 13:183–190

John ER, Prichep LS, Alper KR, Mas FG, Cancro R, Easton P, Sverdlov L (1994) Quantitative electrophysiological characteristics and subtyping of schizophrenia. Biol Psychiatry 36:801–826

Kalischweski P, Schneider D, Kühn HJ (1997) Continuous filtration of cerebrospinal fluid (CSF filtration) in fulminant bacterial meningitis – a new adjuvant therapeutic strategy? In: Gullo A (ed) Anaesthesia pain intensive care and emerging medicine, vol 11. Springer, Berlin Heidelberg New York Tokyo, pp 487–496

Köller H, Siebler M, Hartung HP (1997) Immunologically induced electrophysiological dysfunction: implications for inflammatory diseases of the CNS and the PNS. Progr Neurobiol 52:1–26

Lieb K, Hallensleben W, Czygan M, Stitz L, Staeheli P, and the Bornavirus Study Group (1997) No Borna disease virus specific RNA detected in blood from psychiatric patients in different regions of Germany. Lancet 350:1002

Maes M, Meltzer HY, Bosmans E (1994) Immune-inflammatory markers in schizophrenia: comparison to normal controls and effects of clozapine. Acta Psychiatr Scand 89:346–351

Müller N, Ackenheil M (1995) Immunoglobulin and albumin contents of cerebrospinal fluid in schizophrenic patients: the relationship to negative symptomatology. Schizophr Res 14:223–228

Nair TR, Christensen JD, Kingsbury SJ, Kumar NG, Terry WM, Garver DL (1997) Progression of cerebroventricular enlargement and the subtyping of schizophrenia. Psychiatry Res Neuroimaging 74:141–150

Oxenstierna G, Bergstrand G, Edman G, Flyckt L, Nybäck H, Sedvall G (1996) Increased frequency of aberrant CSF circulation in schizophrenic patients compared to healthy volunteers. Eur Psychiatry 11:16–20

Pfausler B, Grubwieser G, Bösch S, Vollert H, Herold M, Schmutzhard E (1995) Cerebrospinal fluid-filtration reduces TNF alpha in bacterial meningitis CSF. Eur J Neurol 2:570–572

Reiber H, Felgenhauer K (1987) Protein transfer at the blood cerebrospinal fluid barrier and the quantitation of the humoral immune response within the central nervous system. Clin Chim Acta 163:319–328

Richt JA, Pfeuffer I, Christ M, Frese K, Bechter K, Herzog S (1997a) Borna disease virus infection in animals and humans. Em Infect Dis 3:343–352

Richt JA, Alexander RC, Herzog S, Hooper DC, Kean R, Spitsin S, Bechter K, Schüttler R, Feldmann H, Heiske A, Fu ZF, Dietzschold B, Rott R, Koprowski H (1997b) Failure to detect Borna disease virus infection in peripheral blood leukocytes from humans with psychiatric disorders. J Neurovirol 3:174–178

Rother S, Knoblauch KD, Kirschfink M (1995) Filtration of liquor cerebrospinalis (CSF-filtration): technical concept and filter performance under in vitro conditions. Neuropsychiatrie 9:82–85

Rott R, Becht H (1995) Natural and experimental borna disease in animals. In: Koprowski H, Lipkin WI (eds) Borna disease. Springer, Berlin Heidelberg New York Tokyo pp 17–30 (Curr Top Microbiol Immunol)

Rott R, Herzog S, Fleischer B, Winokur A, Amsterdam J, Dyson W, Koprowski H (1985) Detection of serum-antibodies to Borna disease virus in patients with psychiatric disorders. Science 228:755–756

Rott R, Herzog S, Bechter K, Frese K (1991) Borna disease, a possible hazard for man? Arch Virol 118:143–149

Rott O, Herzog S, Cash E (1993) T cell memory specific for self and non-self antigens in rats persistently infected with Borna disease virus. Clin Exp Immunol 93:370–376

Salvatore M, Morzunow S, Schwemmle M, Lipkin WI and the Bornavirus Study Group (1997) Borna disease virus in brains of North American and European people with schizophrenia and bipolar disorder. Lancet 349:1813–1814

Sauder C, Müller A, Cubitt B, Mayer J, Steinmetz J, Trabert W, Ziegler B, Wanke K, Mueller-Lantzsch N, De la Torre JC, Grässer FA (1996) Detection of Borna disease virus (BDV) antibodies and BDV RNA in psychiatric patients: evidence for high sequence conservation of human blood-derived BDV RNA. J Virol 70:7713–7724

Schneider PA, Briese T, Zimmermann W, Ludwig H, Lipkin WI (1994) Sequence conservation in field and experimental isolates of Borna disease virus. J Virol 68:63–68

Sivieri S, Ferrarini AM, Lolli F, Matà S, Pinto F, Tavolato B, Gallo P (1997) Cytokine pattern in the cerebrospinal fluid from patients with GBS and CIDP. J Neurol Sci 147:93–95

Stark DD, Bradley WG Jr (1992) Magnetic resonance imaging, vol 1, 2nd edn. Morley Year Book, St. Louis

Torrey EF (1986) Functional psychoses and viral encephalitis. Integr Psychiatry 4:224–236 (including commentaries)

Torrey EF, Bowler AE, Taylor EH, Gottesman II (1994) Schizophrenia and manic-depressive disorder. Basic Books, New York

Van Os J, Wright P, Murray RM (1997) Follow-up studies of schizophrenia I. Natural history and non-psychopathological predictors of outcome. Eur Psychiatry 12 [Suppl 5]:327s–341s

Waltrip RW II, Buchanan RW, Summerfelt A, Breier A, Carpenter WT Jr, Bryant LN, Rubin SA, Carbone KM (1995) Borna disease virus and schizophrenia. Psychiatry Res 56:33–44

Waltrip RW II, Buchanan RW, Carpenter WT Jr, Kirkpatrick B, Summerfelt A, Breier A, Rubin SA, Carbone KM (1997) Borna disease virus antibodies and the deficit syndrome of schizophrenia. Schizophr Res 23:253–257

Wollinsky KH, Hülser PJ, Brinkmeier H, Mehrkens HH, Kornhuber HH, Rüdel R (1994) Filtration of cerebrospinal fluid in acute inflammatory demyelinating polyneuropathy (Guillain-Barré syndrome). Ann Med Interne 145:451–458

Correspondence: PD Dr. K. Bechter, Department of Psychiatry, Bezirkskrankenhaus Günzburg, Ludwig-Heilmeyer-Strasse 2, D-89312 Günzburg, Federal Republic of Germany

Borna disease virus in patients with psychiatric disorders or fibromyalgia – demonstration of antibodies is highly dependent on the detection method

A. Batra[2], **A. Batra**[1], **K. Schott**[1], and **L. Stitz**[2]

[1]Department of Psychiatry and Psychotherapy, University of Tübingen, and
[2]Federal Research Center for Virus Diseases of Animals, Tübingen,
Federal Republic of Germany

Summary

In the past a possible association between Borna disease virus (BDV)-infection and psychiatric disorders in man was proposed. We investigated sera from patients with schizophrenia, major depression and fibromyalgia for the presence of BDV-specific antibodies and examined the sensitivity of different detection methods for virus specific antibodies. Performing Western blot analysis, immunofluorescence assay and virus neutralisation assay we received inconsistent results. We found a high prevalence (10%) of BDV-specific antibodies in the sera of schizophrenic patients using Western blot, but could not confirme these data by immunofluorescence or virus neutralisation assay. These inconsistent results indicate the need for a critical review of all reported results of significantly enhanced seroprevalence of BDV antibodies in neuropsychiatric patients.

Introduction

Borna disease virus is a negative-sense, single-stranded virus that occurs naturally in a variety of species. This highly neurotropic RNA virus causes immunopathological encephalomyelitis in a variety of infected animals. The resulting Borna disease was first described in horses and sheep, but the natural hosts also include rabbits (Metzler et al., 1978), donkeys (Bilzer et al., 1995), domestic cats (Nowotny and Weissenböck, 1995), cattle (Caplazi et al., 1994) and ostriches (Malkinson et al., 1995).

Infected animals differ in the clinical manifestation of the resulting Borna disease. Most common are disorders of cognition and disturbances of locomotion that can result in paralysis as described in horses, sheep, rabbits and rhesus monkeys (Schmidt, 1912; Zwick and Seifried, 1925; Rott and Becht, 1995).

In experimentally infected rats, antibodies directed against the viral 24 kDa and 38/40 kDa proteins appear about two weeks after infection (p.i.). BDV-specific antibodies with neutralising activity cannot be detected prior to 60 to 80 days p.i.

Because BDV has been also transmitted to primate species, and since the disease in these hosts results in changes of social behaviour (Sprankel et al., 1978; Stitz et al., 1980) speculations concerning the possibility of infections in humans arose.

The fact that BDV shows a strong tropism for the limbic system (Narayan et al., 1983) and excitatory fields in the hippocampus (Gosztonyi and Ludwig, 1984) might be of etiologic relevance in schizophrenia.

Different research groups have screened sera and blood mononuclear cells of humans in recent years. Especially patients with neuropsychiatric disorders such as schizophrenia or major depression (Bechter et al., 1992; Fu et al., 1993; Waltrip et al., 1995; Sauder et al., 1996) but also individuals suffering from chronic fatigue syndrome have been investigated (Kitani et al., 1996) (Table 1).

Sauder et al. (1996) and Bechter and colleagues (1992) investigated the frequency of Borna disease virus specific antibodies in psychiatric patients in general. They found prevalence rates of nearly 6% and 9.6% respectively, whereas controls with no psychiatric disorder showed antibodies in 3.5% only. Sauder et al. (1996) found Borna disease virus RNA in peripheral mononuclear blood cells as well. Waltrip et al. (1995) reported more than 14% antibody-positive schizophrenics, but patients suffering from affective disorders had antibodies in 6.5% only, as shown by Fu and colleagues (1993). Only Deuschle et al. (1997) investigated BDV antigen in cerebrospinal fluid. They found antibodies in no patients with schizophrenia or bipolar disorder but in 9.4% of those suffering from recurrent depression.

Table 1. Frequency of Borna disease virus antibodies in psychiatric patients

Reference	Number of patients	Seroprevalence	Diagnoses
Waltrip et al. (1995)	90	14.4%	schizophrenia
Sauder et al. (1996)	416	9.6%	all psychiatric patients
Fu et al. (1993)	138	9.6%	affective disorders
Bechter et al. (1992)	2377	5.9%	all psychiatric patients
Amsterdam et al. (1985)	265	4.5%	major depression
Rott and Herzog (1985)	979	1.6%	all psychiatric patients
Bechter et al. (1992)	569	3.5%	controls

Purpose

The purpose of our study was to investigate the seroprevalence of BDV-specific antibodies in patients with schizophrenia, major depression and fibromyalgia. By using different techniques we tried to screen for antibodies directed against different viral proteins. To examine the sensitivity of different detection methods for BDV-specific antibodies in psychiatric patients in addition we included cases of fibromyalgia, a disease which is characterised by widespread pain, extreme painful tender points and soft-signs such as tiredness and depressive symptoms.

Methods

Patients

We investigated 40 depressive patients, 50 patients suffering from a schizophrenic psychosis and 40 patients with the diagnosis of fibromyalgia. The schizophrenics were on average about 28 years old, and the other patients about 40 years (Table 2). There were no significant differences in age between male and female patients.

All patients came from an area known to be endemic for Borna disease in animals.

Detection methods

Western blot analysis combined with the enhanced chemiluminescence method was used to screen for the 24 kDa and the 38/40 kDa protein. E. coli expressed recombinant r38 kDa and r24 kDa protein were used as antigens. Polyacrylamid gel electrophoresis was carried out using a 15% acrylamid gel in reducing conditions. The proteins were blotted over night on nitrocellulose membranes. The membranes were blocked with a solution of PBS/0.05% Tween with 10% powdered milk. The sera and the peroxidase conjugated

Table 2. Age and gender

Diagnosis	Sex	N	Age (mean)	Age (Std.dev.)
Major depression	male	16	40.8	9.6
	female	24	41.3	10.6
Schizophrenic psychosis	male	32	28.2	6.4
	female	18	28.3	3.4
Fibromyalgia	male	7	40.7	8.7
	female	33	42.4	7.4

second antibodies were diluted in PBS/0.05% Tween with 5% powdered milk. To visualise bound antibodies, enhanced chemiluminescence (ECL, Amersham) was used. To exclude unspecific cross-reacting of non-specific antibodies with the recombinant proteins, the nitro-cellulose membranes were washed with 7 molar urea after incubation with the diluted sera.

Neutralisation assays were used to detect antibodies with neutralising activities.

The virus neutralisation assay was performed with a modification of the method of Battegay et al. (1991). In 96-well plates 50 ffu of BDV was incubated with dilutions of sera for $1^1/_2$ h at 37°C. Recloned CRL1405 cells (fetal carcass, chemically transformed, guinea pig) were added and after 3 hours of incubation 100 µl of a 1:1 mixture of 2% methylcellulose in double distilled water and 2 × DMEM with 10% FCS was added. After 7 days of incubation at 37°C, infected cells were visualised by intracellular staining of BDV-specific proteins. The dilution of serum necessary to reduce the number of ffu by 50% was defined as the neutralisation titre (NT_{50}). Serum tested in neutralisation assays was heat inactivated at 56°C for 30 minutes.

In addition immunofluorescence assays were carried out, since this method had been used in the majority of studies to date to detect BDV-specific antibodies. The samples were diluted 1:20 in phosphate buffered saline (PBS) and applied to BDV-infected Madine Darby canine kidney (MDCK) cells and uninfected cells of the same type. A biotinylated goat-anti human IgG antibody was used as secondary antibody. In a third step the slides were incubated with Streptavidin conjugated Fluorescein isothyocyanate.

Questionable reactions were retested at dilutions of 1:100, 1:10 and neat.

Results

We found a high prevalence of BDV-specific antibodies in sera from schizophrenic patients using Western blot: 10% of the schizophrenic patients and 5% of the patients with fibromyalgia but only 2.5% of the patients with depression showed antibodies against the viral 24 kDa protein after *multiple* testing. Due to the small sample sizes, no significant differences could be found. No reaction was found against another major virus-specific protein (p38/p40). In virus neutralisation assays 7.5% of the sera from schizophrenics and 5% of the sera from depressives inhibited infection of cells. However, these results could not be reproduced after re-testing (Table 3).

In fact, these findings could not be confirmed by immunofluorescence. Additionally, there were no differences according to gender.

Discussion

Many authors speculate on a role of BDV in the aetiology of depression or other neuropsychiatric disorders. Although some investigators

Table 3. Frequency of Borna disease virus-specific antibodies in psychiatric patients

Diagnosis	N	WB p24	WB p38/40	NT	IF
Major depression	40	2.5%	0%	5%	0%
Schizophrenic psychosis	50	10%	0%	7.5%	0%
Fibromyalgia	40	5%	0%	2.5%	0%

CHI^2 = 2.30; DF = 2, n.s., *WB* Western Blot, *NT* Neutralisation Assay, *IF* Immunfluorescence

believe that BDV infections could be responsible for depression (Bode et al., 1997) or schizophrenia in humans, there has been no proof of this hypothesis yet. Bode and co-workers (1997) reported a single case of infection with Borna disease virus in an elderly woman with a depressive episode and symptoms of Parkinson's disease who was treated successfully with amantadine. Both depression and virus RNA disappeared after a few days of treatment. The authors concluded that amantadine was effective because of its antiviral capabilities. Criticism has to concentrate on the following points:

1. Amantadine may be antidepressive itself.
2. There is no convincing evidence from the literature that amantadine has an anti-BDV effect.
3. Methods of antibody and RNA detection are often not reproducible.

Our research into the prevalence of Borna disease virus antibodies in psychiatric patients shows inconsistent results using different investigation techniques. In particular none of the sera positive for the 24 kDa protein recognised the 38/40 kDa protein in Western blot. In experimentally infected animals antibodies specific for both these proteins occur in detectable levels at an early stage after infection.

Furthermore none of the sera positive for the 24 kDa protein showed a positive reaction in immunofluorescence assay. Though evaluation of immunofluorescence assays is very subjective these assays have the advantage, of screening for antibodies specific for all viral proteins. However, Bechter and colleagues (1987) reported that immunofluorescence assays performed with low-titre sera are unreliable in the testing of human sera for BDV-specific antibodies.

Besides this, none of the sera positive in Western blot showed neutralising activity. In experimentally infected rats neutralising antibodies occur in a late stage after infection during the chronic phase of the disease. Therefore we conclude that all results reported in the literature have to be interpreted with caution, although the schizophrenics tested in this research show highest prevalence in accordance with the published data. The inconsistent findings indicate the need for a critical review of all reported results of significantly enhanced seroprevalence of BDV antibodies in neuropsychiatric patients, especially if only one single test method was employed.

References

Amsterdam JD, Winokur A, Dyson W, Herzog S, Gonzalez F, Rott R, Koprowski H (1985) Borna disease virus a possible etiologic factor in human affective disorders? Arch Gen Psychiatry 42:1093–1096

Battegay M, Cooper S, Althage A, Bänzinger J, Hengartner H, Zinkernagel RM (1991) Quantification of lymphocytic choriomeningitis virus with an immunological focus assay in 24- or 96-well plates. J Virol Meth 33:191–198

Bechter K, Herzog S, Fleischer B, Schüttler R, Rott R (1987) Kernspintomographische Befunde bei psychiatrischen Patienten mit und ohne Serum-Antikörper gegen das Virus der Bornaschen Krankheit. Nervenarzt 58:617–624

Bilzer T, Stitz L (1994) Immune-mediated brain atrophy. J Immunol 153:818–823

Bilzer T, Planz O, Lipkin WI, Stitz L (1995) Presence of CD4+ and CD8+ T cells and expression of MHC class I and MHC class II antigen in horses with Borna disease virus-induced encephalitis. Brain Pathol 5:223

Bode L, Dietrich DE, Stoyloff R, Emrich HM, Ludwig H (1997) Amantadine and human Borna disease virus in vitro and in vivo in an infected patient with bipolar depression [letter]. Lancet 349:178–179

Caplazi P, Waldvogel A, Stitz L, Braun U, Ehrensperger F (1994) Borna disease in naturally infected cattle. J Comp Path 111:65–72

Deuschle M, Heuser I, Gotthardt U, Schmider J, Bode L, Ludwig H (1997) Borna disease virus antigen in cerebrospinal fluid of patients with recurrent depression and multiple sclerosis [abstract]. Pharmacopsychiat 30: 160

Fu ZF, Amsterdam JD, Kao M, Shankar V, Koprowski H, Dietzschold B (1993) Detection of Borna disease virus-reactive antibodies from patients with affective disorders by western immunoblot technique. J Affect Disord 27:61–68

Kitani T, Kuratsune H, Fuke I, Nakamura Y, Nakaya T, Asahi S, Tobiume M, Yamaguti K, Machii T, Inagi R, Yamanishi K, Ikuta K (1996) Possible correlation between Borna disease virus infection and Japanese patients with chronic fatigue syndrome. Microbiol Immunol 40:459–462

Malkinson M, Weisman Y, Perl S, Ashash E (1995) A Borna-like disease of ostriches in israel. Curr Top Microbiol Immunol 190:31

Metzler A, Ehrensperger F, Wyler R (1978) Natürliche Bornavirus Infektion bei Kaninchen. Zbl Vet Med B 25:161

Nowotny N, Weissenböck H (1995) Description of feline nonsuppurative meningoencephalomyelitis ("staggering disease") and studies of its etiology. J Clin Microbiol 33:1668

Rott R, Gerzog S (1985) Detection of serum antibodies to Borna disease virus in patients with psychiatric disorders. Science 228:755–756

Sauder C, Muller A, Cubitt B, Mayer J, Steinmetz J, Trabert W, Ziegler B, Wanke K, Mueller Lantzsch N, de la Torre JC, Grasser FA (1996) Detection of Borna disease virus (BDV) antibodies and BDV RNA in psychiatric patients: evidence for high sequence conservation of human blood-derived BDV RNA. J Virol 70:7713–7724

Schmidt J (1912) Untersuchungen über das klinische Verhalten der seuchenhaften Gehirn-Rückenmarksentzündung (Bornasche Krankheit) des Pferdes nebst Angaben über diesbezügliche therapeutische Versuche. Berliner tierärztl Wschr 28:581–597

Sprankel H, Richards K, Ludwig H, Rott R (1978) Behavior alterations in tree shrews (Tupaia glis) induced by Borna disease virus. Med Microbiol Immunol 165:1

Stitz L, Krey H, Ludwig H (1980) Borna disease in rhesus monkeys as a model for uevo-cerebral symptoms. J Med Virol 6:333

Waltrip RW, Buchanan RW, Summerfelt A, Breier A, Carpenter WT, Jr, Bryant NL, Rubin SA, Carbone KM (1995) Borna disease virus and schizophrenia. Psychiatry Res 56:33–44

Zwick W, Seifried O (1925) Übertragbarkeit der seuchenhaften Gehirn-Rückenmarksentzündung des Pferdes (Bornasche Krankheit) auf kleine Versuchstiere (Kaninchen). Berliner Tierärztl Wschr 41:129

Correspondence: Dr. A. Batra, University of Tübingen, Department of Psychiatry and Psychotherapy, Osianderstrasse 24, D – 72076 Tübingen, Federal Republic of Germany

Interaction of the serotonergic and the immune systems: 5HT-Moduline

B. Grimaldi, A. Bonnin, M.P. Fillion, O. Massot, J.C. Rousselle, N. Prudhomme, and **G. Fillion**

Unité de Pharmacologie-Neuro-Immuno-Endocrinienne, Institut Pasteur, Paris, France

Summary

The serotonergic system is known as a neuromodulatory system present in the brain and participating to the homeostasis of the brain. Its major morphological and anatomical features are favoring this role. Accordingly it is involved in many physiological functions and a number of psychiatric dysfunctions. Amongst the numerous receptors involved in its functions, $5HT_{1B}$ receptors constitute a particular subtype located on neuron terminals and regulating the release of the corresponding neurotransmitter. These receptors are specifically the target of an endogenous tetrapeptide (5HT-Moduline) characterized in mammalian brain and regulating the functional activity of the receptor as an allosteric modulator.

Immunocompetent cells were shown to also contain $5HT_{1B}$ receptors using molecular biology, immunocytochemistry and pharmacology. These receptors participate to the control of transcriptional activity of immediate early genes, and stimulate the proliferation of cells. 5HT-Moduline also interacts with the function of $5HT_{1B}$ receptor at this level, playing an antagonistic role.

The observed results indicate that the serotonergic system not only exerts a neuromodulatory role in central nervous tissue but presumably also controls the activity of immunocompetent cells. The serotonergic system may be involved in the reciprocal neuroimmune relationships via mechanisms which implicate $5HT_{1B}$ receptors and their allosteric modulator 5HT-Moduline.

The serotonergic system in the central nervous system

The neuroanatomy and the neuromodulatory role of the serotonergic system

The serotonergic system is largely considered as a neuromodulatory system which interacts with other neurotransmissions in the brain and participates to the elaboration of an adapted response of the central nervous system (CNS) to external stimuli.

This system is strictly centralized, the cellular bodies at the origin of the central serotonergic innervation being essentially localized in the raphe nuclei. It is very ubiquitous, the 5HT neurons projecting to a vast majority of the brain areas (Jacobs and Azmitia, 1992; Baumgarten and Grozdanovic, 1994, 1997).

Although the number of serotonergic cells in the brain is limited (11500 and 24000 respectively in rat and cat raphe dorsalis) (Wiklund et al., 1981) the serotonergic system possesses the property and the capacity to interact with other neurons. Indeed two main types of 5HT neurons have been described in the brain. It is the classical schematic representation of thick 5HT neurons functioning with a mode of action corresponding to a synaptic transmission; these neurons interact only with a few other neurons. Moreover, the system developed smaller arborescent axonal projections (Kosofsky and Molliver, 1987; Takeuchi, 1988) possessing numerous varicosities along their axons equivalent of neuron terminals. It was estimated that a single axon projecting from the raphe to the cortex in human brain could contain up to 500,000 varicosities and that in the rat brain, the density of the terminals in the cortex could reach 6×10^6 per mm^3 of the tissue (Audet et al., 1989). A high proportion of these varicosities are non-junctional (Descarries et al., 1990) with a mode of so called volumic transmission (Fuxe et al., 1989) permitting the diffusion of 5HT in the extracellular medium for an action at distance on specific receptors and favouring the capacity for interaction of a single serotonergic fiber with other neurons.

Thus, the morphological characteristics of the 5HT system are in agreement with its modulatory role since this system is very centralized, ubiquitous and exhibits a high number of varicosities for a high capacity of communication.

It is now clearly established that serotonin is involved in a large number of physiological functions as temperature regulation, sleep, learning and memory, behaviour, sexual functions, hormonal secretions and immune activity and also in pathological disorders particularly in stress, anxiety and depression (Zifa and Fillion, 1992).

$5HT_{1B}$ receptors in serotonergic activity

At least 14 different types of 5HT receptors mediate the 5HT activity and among them the $5HT_{1B}$ receptors play an important role in the function of the serotonergic system (Hoyer et al., 1994; Hartig, 1997).

The $5HT_{1B}$ receptor subtypes correspond to the $r5HT_{1B}$ (rat) and $h5HT_{1B}$ (human) receptors, previously called $5HT_{1B}$ and $5HT_{1D\beta}$ respectively. These receptors belong to the multigenic family of the G-protein coupled receptors having a monomeric structure with seven highly conserved transmembrane domains. We distinguish two categories of receptors: autoreceptors and heteroreceptors. Autoreceptors are localized on 5HT varicosities where they inhibit the evoked release of 5HT and its biosynthesis (Middlemiss, 1984; Engel et al., 1986; Fink et al., 1995; Hjorth et al., 1995) and also in the raphe on somatodendritic fibers of the serotonergic system (Piñeyro et al., 1995; Piñeyro and Blier, 1996). Heteroreceptors are located on non-serotonergic terminals where they inhibit the corresponding neurotransmitters release. In sum, the role of the $5HT_{1B}$ autoreceptors is to inhibit the release of 5HT to finely modulate the tonic serotonergic activity.

Characterization and biochemical and pharmacological properties of 5HT-Moduline

Few years ago, direct and non-competitive interactions of antidepressant drugs with $5HT_1$ receptor subtypes have been shown in the laboratory (Fillion and Fillion, 1981) and based on these experimental observations, it was proposed the hypothesis of the existence of an endogenous compound which could directly interact with $5HT_1$ receptor subtypes to affect their activity.

Recently, we isolated and characterized in the laboratory an endogenous peptide from bovine and horse brain extracts (Rousselle et al., 1996). This peptide which we called 5HT-Moduline is a tetrapeptide composed of Leu-Ser-Ala-Leu (LSAL). Then, this peptidic sequence was synthetized and tested in pharmacological studies to determine its properties. Binding experiments have shown that 5HT-Moduline interacts specifically with $5HT_{1B}$ receptors in a non-competitive mechanism with a high apparent affinity suggesting the existence of 2 sites of action on the receptor, one for the 5HT and another one for the peptide (Massot et al., 1996). 5HT-Moduline interacts specifically with the $5HT_{1B}$ receptors and does not interact with other serotonergic and non-serotonergic receptors tested.

The functional activity of 5HT-Moduline was measured on the second messenger system and at cellular level on synaptosomal 5HT release. Indeed, the $5HT_{1B}$ receptor is a G-protein related receptor coupled to an adenylate cyclase; the effect of a $5HT_{1B}$ agonist is to inhibit the adenylate cyclase activity and inhibit the 5HT release. So, it was shown that 5HT-Moduline reverses the inhibition induced by a $5HT_{1B}$ agonist (CP93129) on the adenylate cyclase activity and reverses the inhibitory effect of a $5HT_{1B}$ agonist (CGS12066B) on the synaptosomal 5HT release.

Furthermore intracerebroventricular injection of 5HT-Moduline completely reverses the action of a selective $5HT_{1B}$ agonist (RU24969) in an *in vivo* model of social interaction (Massot et al., 1996).

Thus, these results indicate that 5HT-Moduline specifically interacts with $5HT_{1B}$ receptors with a non-competitive interaction in binding experiments and with an antagonistic effect in *in vitro* and *in vivo* functional studies.

Distribution of 5HT-Moduline in the brain

Then, polyclonal antibodies against 5HT-Moduline have been raised in the laboratory to specifically determine the immunocytochemical localization of neurons expressing the peptide in the mouse brain (Grimaldi et al., 1997). The general distribution of the 5HT-Moduline immunoreactivity appeared heterogenous in the mouse brain and suggests the presence of the peptide in cortical and limbic areas as well as in basal ganglia. These results also indicate that 5HT-Moduline is not a circulating substance since it is shown to be present in cellular structures probably neuronal cell bodies with axon-like processes suggesting that the peptide is localized in neuronal cells (Grimaldi et al., 1997).

Accordingly, it was shown in *in vitro* experiments that 5HT-Moduline was released from rat brain synaptosomes in a Ca^{++}/K^{+} dependent manner as a neurotransmitter (Massot et al., 1996).

Recently, using sequence homology, a gene coding for a novel chromogranin called NESP55 was cloned (Ischia et al., 1997). The corresponding protein is likely the precursor of 5HT-Moduline since the sequence of the peptide surrounded with basic residues, well known to be cleavage sites, is contained in the protein sequence. Therefore, 5HT-Moduline is potentially processed from this protein.

The role of $5HT_{1B}$ receptors and 5HT-Moduline in physiopathology

The involvement of 5HT-Moduline in physiopathology was also studied. Indeed, the increase of 5HT release in stress situations which was previously proposed (Briley et al., 1990) is yet demonstrated using microdialysis and tissue punch extraction techniques (Shimiju et al., 1992; Kahawara et al., 1993; Zangen et al., 1997). It was hypothesized that this phenomenon may involve $5HT_{1B}$ receptors. In the laboratory, it was shown that rats subjected to an acute stress session as a restraint stress in a glass tube for 40 min, exhibited a marked decrease of the $5HT_{1B}$ receptor activity in substantia nigra (Bolaños-Jimenez et al., 1995). These results indicated that an acute stress of immobilization induced the desensitization of $5HT_{1B}$ receptors and that this effect occurred early after stress. It could be hypothesized that 5HT-Moduline was released during the stress and contributed to the observed desensitization. To further test this hypothesis, the effect of intracerebral injections of the synthetic peptide, the scrambled peptide (ALLS) and

NaCl were studied on the adenylyl cyclase activity stimulated by $5HT_{1B}$ receptors agonists in the rat substantia nigra. The results have shown that the peptide had a similar desensitizing effect of the autoreceptor as that described after a stressful situation (Seguin et al., 1997). These observations suggested that acute stress may induce the release of 5HT-Moduline, i.e. the peptide desensitizes $5HT_{1B}$ receptors leading to an increase in 5HT activity. To validate this hypothesis we examined the kinetic of 5HT-Moduline production following a stress of immobilization. Using 5HT-Moduline antibodies, we observed that 5HT-Moduline levels markedly increased in various regions of rat brain i.e. cortex, hippocampus, hypothalamus, substantia nigra whereas it was not significantly modified in striatum. The maximal increase in 5HT-Moduline occurred 15 min to 25 min after stress depending of the brain region and return to basal level occurred during the first hour following stress (Bonnin et al., 1999).

These results indicate the existence of an efficient mechanism rapidly modulating the serotonergic activity in the central nervous system, conferring to the serotonergic system the capacity of tuning finely and differentially other neurotransmissions in the brain.

The serotonergic system in the immune system

In the central nervous system, a regulatory loop involving 5HT, $5HT_{1B}$ receptors and 5HT-Moduline is observed. It is of interest to examine whether this regulatory element also exists at the periphery and would be potentially playing a role in the interactions between the central nervous system and the immune system (IS).

The neuroimmune interactions

The classical ways of communication between the central and the immune system, with the main protagonists as the autonomic nervous system, the endocrine system, the lymphokines and cytokines systems, (Blalock, 1989; Rabin et al., 1989; Chaouloff, 1993; Mc Ewen et al., 1997; Walter, 1997; Weigent and Blalock, 1997) indicated that for each of these systems, specific receptors i.e. adrenergic, noradrenergic, corticosteroid, lymphokine or cytokine receptors are present as on the immune cells for direct actions as they are on the brain cells (Stevens-Felten and Bellinger, 1997; Garza and Carr, 1997).

The hypothesis of the existence of common receptors and mediators for central and peripheral systems constituted the basis of our experimental work. We examined the involvement of the serotonergic system in the CNS-IS interactions and focused our investigations on the role of 5HT, $5HT_{1B}$ receptors and 5HT-Moduline in mechanisms involved in the regulation of the IS function.

In order to validate this model, the experiments were carried out in rat and mouse on thymocytes and splenocytes, in human on lymphocytes and using a human T lymphoblastoid cell line, the CEM.

$5HT_{1B}$ receptors in immunocompetent cells

We have shown that $5HT_{1B}$ receptors are present on immunocompetent cells by RT-PCR amplification, autoradiographic binding and immunocytochemical studies.

The presence of mRNAs specifically coding for the $5HT_{1B}$ receptors have been detected in the immunocompetent cells using RT-PCR amplification. Then, the existence of high affinity binding sites for 5HT was demonstrated using autoradiographic binding on rat spleen as well as on human leucocytes. The presence of $5HT_{1B}$ receptors also was demonstrated on CEM by immunocytochemical studies using $5HT_{1B}$ receptor antibodies raised in the laboratory (Grimaldi et al., 1995).

The obtained results demonstrated that $5HT_{1B}$ receptors are present on immunocompetent cells (Grimaldi et al., 1998). As yet, only one gene coding for $5HT_{1B}$ receptor has been cloned; therefore, the same $5HT_{1B}$ receptor likely exists on neuron terminals and on immunocytes.

Functional activity of $5HT_{1B}$ receptors in immunocompetent cells

The functional activity of $5HT_{1B}$ receptors was examined on immune cells by measuring their capacity to induce the transcriptional activity of various immediate early genes (IEG). Indeed, IEG are characterized by their rapid and transient induction in response to a wide variety of extracellular stimuli. These genes encode proteins involved as transcription factors regulating the expression of other target genes playing a role in various cellular mechanisms. We observed that specific $5HT_{1B}$ receptors activation induces a transitory increase of JunD mRNA expression in spleen and thymus of mouse and in CEM. Indeed the JunD mRNA expression is rapidly and specifically induced by a $5HT_{1B}$ agonist, CP93129 in rat and mouse immunocompetent tissues and L694247 in human T cells.

Therefore, these results indicate that $5HT_{1B}$ receptors are present on immunocompetent cells and are functional, since they specifically induce IEG mRNA expression (Grimaldi et al., 1998).

Role of $5HT_{1B}$ receptors in immunocompetent cells

Cellular proliferation of immunocompetent cells was studied to determine the physiological consequences of the serotonergic activity.

A pharmacological study on cell proliferation revealed that $5HT_{1B}$ receptors present on the CEM specifically mediated an increase of their proliferation. Then, we demonstrated on CEM cells that 5HT-Moduline inhibited the proliferative effect induced by a $5HT_{1B}$ agonist, L694247.

Thus, the results demonstrate that 5HT-Moduline has an antagonistic effect on the proliferative action of a $5HT_{1B}$ agonist on immunocompetent cells. A similar antagonistic effect was already observed in the central nervous system (Grimaldi et al., 1998).

Conclusion

In conclusion, the results obtained in these experiments indicate that $5HT_{1B}$ receptors are present on immunocompetent cells and they exhibit functional activity. Moreover it appears that 5HT-Moduline has the ability to interact with the activity of the immune system and thus, it may exert a role in the regulation of the immune response as it appears to do at the central nervous system.

The existence of the potential precursor gene of the 5HT-Moduline and its presence in the adrenal medulla (Ischia et al., 1997) suggests that the immune system may possess similarities with the central nervous system and may constitute a model for the study of the involvement of 5HT-Moduline in numerous physiological as well as pathological events as stress or immune related disorders.

Moreover, central pathological dysfunctions as stress, anxiety or depression in which $5HT_{1B}$ receptors and 5HT-Moduline are involved may produce immune disorders as immunodepression (Chaouloff, 1993; Dantzer and Mormede, 1995; Rosch, 1995). Reciprocally, it may be proposed that immune disorders involving $5HT_{1B}$ and 5HT-Moduline may be related to psychiatric diseases involving the serotonergic system as the result of reciprocal interactions between the central nervous system and the immune system.

References

Audet M, Descarries L, Doucet G (1989) Quantified regional and laminar distribution of the serotonin innervation in the anterior half of the adult cerebral cortex. J Chem Neuroanat 2:29–44

Baumgarten H, Grozdanovic Z (1994) Neuroanatomy and neuro-physiology of central serotonergic systems. J Serotonin Res 3:171–179

Baumgarten HG, Grozdanovic Z (1997) Anatomy of central serotoninergic projection system. Handb Exp Pharmacol 129:42–89

Blalock JE (1989) A molecular basis for bidirectional communication between the immune and neuroendocrine systems. Physiol Rev 69:1–32

Bolaños-Jiménez F, Manhaes de Castro R, Seguin L, Cloëz-Tayarani I, Monneret V, Drieu K, Fillion G (1995) Effects of stress on the functional properties of pre- and postsynaptic 5-HT_{1B} receptors in the rat brain. Eur J Pharmacol 294:531–540

Bonnin A, Grimaldi B, Fillion MP, Fillion G (1999) Acute stress induces a differential increase of 5HT-Moduline tissue content in various rat brain areas. Brain Res (in press)

Briley M, Chopin P, Moret C (1990) Effect of serotonergic lesion on "anxious" behaviour measured in the elevated plus-maze test in the rat. Psychopharmacoloy 101:187–189

Choualoff F (1993) Physiopharmacological interactions between stress hormones and central serotoninergic system. Brain Res Rev 18:1–32

Dantzer R, Mormede P (1995) Psychoneuroimmunology of stress. In: Leonard BE, Miller K (eds) Stress, the immune system and psychiatry. Wiley, Chichester, pp 47–67

Descarries L, Audet MA, Doucet G, et al. (1990) Morphology of central serotonin neurons. In: Whitaker-Azmitia PM, Peroutka SJ (eds) The neuropharmacology of serotonin, vol 600. Academic Press, New York, pp 81–92

Engel G, Göthert M, Hoyer D, Schlicker E, Hillenbrand K (1986) Identity of inhibitory presynaptic 5-hydroxytryptamine (5-HT) autoreceptors in the rat brain cortex with 5-HT_{1B} binding sites. Naunyn Schmiedebergs Arch Pharmacol 332:1–7

Fillion G, Fillion M (1981) Modulation of affinity of postsynaptic serotonin receptors by antidepressant drugs. Nature 292:349–351

Fink K, Zenter J, Göthert M (1995) Subclassification of presynaptic 5-HT autoreceptors in the human cerebral cortex as $5\text{-HT}_{1D\beta}$ receptors. Naunyn Schmiedebergs Arch Pharmacol 352:451–454

Fuxe K, Von Euler G, Agnati LF, Ögren S-O (1989) Galanin selectively modulates 5-hydroxytryptamine 1A receptors in the rat ventral limbic cortex. Neurosci Lett 85:163–167

Garza Jr HH, Carr DJJ (1997) Neuroendocrine peptide receptors on cells of the immune system. In: Blalock JE (ed) Neuroimmunoendocrinology, vol 69. Birmingham, Ala, pp 132–154

Grimaldi B, Cloez I, Fillion MP, Mazié JC, Hen R, Fillion G (1995) Production and characterization of an antibody directed against the mouse 5-HT_{1B} receptor. Neurosci Res 24:97–101

Grimaldi B, Fillion MP, Bonnin A, Rousselle O, Massot O, Fillion G (1997) Immunocytochemical localization of neurons expressing 5-HT moduline in the mouse brain. Neuropharmacology 36:1079–1087

Grimaldi B, Sibella-Arguelles C, Bonnin A, Fillion MP, Massot O, Rousselle JC, Prudhomme N, Fillion G (1998) Control of the serotonergic activity and gene expression by 5-HT-moduline. 11th ECNP Congress, Paris

Hartig PR (1997) Molecular biology and tranductional characteristics of 5HT receptors. Handb Exp Pharmacol 129:176–212

Hjorth S, Suchowski S, Galloway MP (1995) Evidence for 5-HT autoreceptor-mediated, nerve impulse-independent, control of 5-HT synthesis in the rat brain. Synapse 19:170–6

Hoyer D, Clarke DE, Fozard JR, Hartig PR, Martin GR, Mylecharane EJ, Saxena PR, Humphrey PPA (1994) VII. International union of pharmacology. Classification of receptors for 5-hydroxytryptamine (serotonin). Pharmacol Rev 46:157–203

Ischia R, Lovisetti-Scamihorn P, Hogue-Angeletti R, Wolkersdorfer M, Winkler H, Fisher-Colbrie R (1997) Molecular cloning and characterisation of NESPP55, a novel chromogranin-like precursor of a peptide with 5-HT_{1B} receptor antagonist activity. J Biol Chem 17:11657–11662

Jacobs B, Azmitia E (1992) Structure and function of the brain serotonin system. Pharmacol Rev 72:165–229

Kawahara H, Yoshida M, Yokoo H, Nishi M, Tanaka M (1993) Psychological stress increases serotonin release in the rat amygdala and prefrontal cortex assesses by *in vivo* microdialysis. Neurosci Lett 162:81–84

Kosovfsky BE, Molliver ME (1987) The serotonin innervation of cerebral cortex: different classes of axon terminals arise from dorsal and median raphe nuclei. Synapse 1:153–68

Massot O, Rousselle JC, Fillion MP, Grimaldi B, Cloëz-Tayarani I, Fugelli A, Prudhomme N, Seguin L, Rousseau B, Plantefol M, Hen R, Fillion G (1996) 5-hydroxytryptamine-moduline, a new endogenous cerebral peptide, controls the serotoninergic activity via its specific interaction with 5-hydroxytryptamine 1B/1D receptors. Mol Pharmacol 50:752–762

McEwen BS, Biron CA, Brunson KW, Bulloch K, Chambers WH, Dhabhar FS, Goldfarb RH, Kitson RP, Miller AH, Spencer RL, Weiss JM (1996) The role of adrenocorticoids as modulators of immune function in health and disease: neural, endocrine and immune interactions. Brain Res Rev 23:79–133

Middlemiss DN (1984) Stereoselective blockade at [^{3}H]5-HT binding sites and at the 5-HT autoreceptor by propanolol. Eur J Pharmacol 101:289–293

Pineyro G, Blier P (1996) Regulation of 5-hydroxytryptamine release from rat midbrain raphe nuclei by 5-hydroxytryptamine 1D receptors: effect of tetrodoxin, G protein inactivation and long-term antidepressant administration. J Pharmacol Exp Ther 276:697–707

Pineyro G, Castanon N, Hen R, Blier P (1995) Regulation of [^{3}H]5-HT release in raphe, frontal cortex and hippocampus of 5-HT_{1B} knock-out mice. NeuroReport 7:353–9

Rabin BS, Cohen S, Ganguli R, Lysle DT, Cunnick JE (1989) Bidirectional interaction between the central nervous system and the immune system. Crit Rev Immunol 9:279

Rosh PJ (1995) Future directions in psychoneuroimmunology : psychoelectroneuroimmunology? In: Leonard BE, Miller K (eds) Stress, the immune system and psychiatry. Wiley, Chichester, pp 207–231

Rousselle JC, Massot O, Delepierre M, Zifa E, Fillion G (1996) Isolation and characterization of an endogenous peptide from rat brain interacting specifically with the serotoninergic 1B receptor subtypes. J Biol Chem 271:726–735

Seguin L, Seznec JC, Fillion G (1997) The endogenous cerebral tetrapeptide 5-HT-moduline reduces in vivo the functional activity of central 5-HT_{1B} receptors in the rat. Neurosci Res 27:277–280

Shimizu N, Take S, Hori T, Oomura Y (1992) In vivo measurement of hypothalamic serotonin release by intracerebral microdialysis: significant enhancement by immobilisation stress in rats. Brain Res Bull 28:727–734

Stevens-Felten SY, Bellinger DL (1997) Noradrenergic and peptidergic innervation of lymphoid organs. In: Blalock JE (ed) Neuroimmunoendocrinology, vol 69. Birmingham, Ala, pp 99–131

Takeuchi Y (1988) Distribution of serotonin neurons in the mammalian brain. In: Osborne NN, Hamon M (eds) Neuronal serotonin. John Wiley & Sons, London, pp 25–56

Walter MR (1997) Structural biology of cytokines, their receptors, and signaling complexes: implications for the immune and neuroendocrine circuit. In: Blalock JE (eds) Neuroimmunoendocrinology, vol 69. Birmingham, Ala, pp 76–98

Weigent DA, Blalock JE (1997) Production of peptide hormones and neurotransmitters by the immune system. In: Blalock JE (ed) Neuroimmunoendocrinology, vol 69. Birmingham, Ala, pp 1–30

Wiklund L, Leger L, Persson M (1981) Monoamine cell distribution in the cat brain. A fluorescence histochemical study with quantification of indolaminergic and locus coeruleus cell groups. J Comp Neurol 203:613–47

Zangen A, Overstreet DH, Yadid G (1997) High serotonin and 5-hydroxyindoleacetic acid levels in limbic brain regions in a rat model of depression: normalization by chronic antidepressant treatment. J Neurochem 69(6):2477–2483

Zifa E, Fillion G (1992) 5-hydroxytryptamine receptors. Pharmacol Rev 44:401–458

Correspondence: Dr. B. Grimaldi, Unité de Pharmacologie-Neuro-Immuno-Endocrinienne, Institut Pasteur, 28 rue du docteur Roux, F-75724 Paris cedex, France

The inflammatory response system activation model of major depression

M. Maes

Clinical Research Center for Mental Health, and University Department of Psychiatry, AZ Stuivenberg, Antwerp, Belgium
Department of Psychiatry, Vanderbilt University, Nashville, TN, USA

Summary

The present paper proposes a concise inflammatory response system (IRS)-activation model of major depression. The evidence that alterations in the IRS participate in the pathophysiology or even etiology of major depression consists of the following. 1) Confirmed findings of in vivo IRS activation in major depression, such as increased numbers of leukocytes, monocytes, neutrophils, activated T-lymphocytes, increased secretion of neopterin and prostaglandins, and increased secretion of proinflammatory cytokines, such as interleukin-1β (IL-1β), IL-6 and interferon-γ (IFNγ). 2) A variety of antidepressants, such as selective serotonin reuptake inhibitors and tricyclic antidepressants, have negative immunoregulatory activities. 3) Activation of the IRS in major depression is significantly related to some of the neuroendocrine disorders in that illness. 4) Proinflammatory cytokines, such as IL-1, IL-6 and interferons given to humans or animals can produce depressive symptoms or full blown major depression. 5) The IRS activation model may account for the multicausal etiology of major depression whereby external or psychosocial stressors (negative life events), and internal or organic stressors are considered to play a pivotal role in the etiology of depression.

Introduction

Current models concerning the biological pathophysiology of depression emphasize the role of brain monoaminergic neurotransmitters, such as serotonin (5-HT) and catecholamines, and hypothalamic-pituitary-adrenal (HPA)-axis hyperactivity [1]. The contemporary models of depression do not incorporate the immune system, even though immune functions may powerfully influence behavior, serotonergic and catecholaminergic metabolism in the brain and HPA-axis activity. In the present paper we propose a concise inflammatory response system (IRS)-activation model of major depression. The evidence that alterations in the IRS participate in the pathophysiology or even pathogenesis of major depression consists of the following.

Confirmed findings of increased numbers of leukocytes, monocytes, neutrophils, activated T-lymphocytes, increased secretion of neopterin and prostaglandins

Herbert and Cohen [2] showed, in a meta-analytical review, that major depression is associated with a significant increased number of leukocytes and neutrophils. Increased numbers of immune cells, including monocytes and T helper cells, were reported to occur in depression [3–7]. Other findings include an increased number of activated T cells [8,9] and increased secretion of neopterin, a very sensitive marker of activation of cell-mediated immunity [10–13]; and prostaglandin PGE2 [14–16].

Confirmed findings on increased secretion of proinflammatory cytokines, such as IL-1β, IL-2 (or the IL-2 receptor, IL-2R), IL-6 and interferon-γ (IFNγ)

Increased production of IL-1β, IL-2, IL6 and IFNγ has been reported to be present in depression [6,17–20]. Increased circulating levels of the soluble IL-2-receptor (sIL-2R) point toward T cell activation in depression [6,17,21].

Confirmed findings on the presence of an acute phase response in depression

There are many reports that major depression is characterized by increased plasma concentrations of positive acute phase proteins (APPs), such as haptoglobin (Hp), α1-antitrypsin, ceruloplasmin, α1-acid glycoprotein, C-reactive protein, α1-antichymotrypsin; and decreased plasma concentrations of negative AP proteins, such as albumin and transferrin [22–25]. The chronic mild stress (CMS) model of depression in the rodent and the olfactory bulbectomized (OB) rat model of depression are also characterized by an AP response [26,27]. There is evidence that the AP response in depression is related to increased production of proinflammatory cytokines, such as IL-6 [22].

A variety of tricyclic antidepressants and selective serotonin reuptake inhibitors (SSRIs) have negative immunoregulatory activities

There is now evidence that antidepressants may have suppressive activities on the secretion of proinflammatory cytokines in vivo as well as in vitro and that these drugs can suppress the AP response. Subchronic treatment with fluoxetine has been shown to normalize initially increased serum IL-6 concentrations in depressed patients [28]. Subchronic treatment with tricyclic antidepressants and SSRIs suppress the AP response in major depression [29] and in the olfactory bulbectomized and chronic mild stress models of depression in the rat [26,27]. Xia et al. [30] found that clomipramine, imipramine and citalopram significantly suppressed the secretion of IL-2 by in vitro stimulated T lymphocytes and of IL-1β and TNFα by stimulated monocytes. We found that clomipramine, sertraline and trazodone, at concentrations in the therapeutical range, significantly reduced IFNγ secretion, whereas clomipramine and sertraline significantly increased the secretion of IL-10, a negative immunoregulatory cytokine, in culture supernatants (Maes et al., 1999). All three antidepressants significantly reduced the IFNγ/IL-10 ratio (Maes et al., 1999). The above results suggest that antidepressants have highly specific, negative immunoregulatory effects.

Activation of the IRS in major depression is associated with the neuroendocrine disorders in that illness

The actions of cytokines offer a possible explanation to the serotonergic and HPA-axis dysfunctions in depression. Lower availability of tryptophan to the brain, which is one of the most frequently reported serotonergic abnormalities in depression, is inversely related to various signs of IRS activation in that illness [12,20]. It is well-known that an immune-inflammatory response with increased production of proinflammatory cytokines, is accompanied by decreased plasma total tryptophan. This phenomenon may be caused by i) induction of indoleamine 2,3 dioxygenase (IDO), a major tryptophan catabolizing enzyme, by cytokines, such as IFNγ, IL-1 and IL-2; and ii) muscle breakdown of proteins, increased release of amino acids, inhibited muscle amino acid uptake, and increased use of some amino acids for leukocyte activity and synthesis of acute phase proteins [12,20].

It is equally well established that proinflammatory cytokines, such as IL-1, IL-6 and tumor necrosis factor-α (TNFα) activate the HPA-axis by i) stimulating the release of corticotropin releasing hormone (CRH), adrenocorticotrophic hormone (ACTH) and cortisol [31]; and ii) by decreasing glucocorticoid receptor (GR) expression or GR translocation from the cytoplasm to the nucleus [32]. In depression, there are significant positive correlations between postdexamethasone cortisol or baseline plasma cortisol values and either serum IL-6 concentrations or the secretion of IL-1 and IL-6 by peripheral blood mononuclear cells [31].

Proinflammatory cytokines, such as IL-1, IL-6 and IFN given to humans or animals can produce depressive symptoms and behavior

An AP response not only encompasses a broad array of immune, metabolic and neuroendocrine alterations, but also a specific symptom complex, i.e. sickness behavior, that is psychomotor retardation, irritability, anhedonia, anorexia, weight loss, loss of interest, lethargy and sleep disorders [33]. Proinflammatory cytokines are key mediators of the various components of this behavioral complex [33]. Acute or repeated administration of LPS and IL-1 to the rodent may induce "sickness behavior"[33], while endotoxin administration to rats decreased the free consumption of saccharin, a model of anhedonia or the inability to experience pleasure [34]. Interferons given to humans can produce symptoms and behavior similar, if not identical to those found in depression. These include affective, cognitive and vegetative symptoms, e.g. anhedonia, fatigue, social withdrawal, depressed mood, irritability, lack of interest, psychomotor retardation, poor concentration, sleep disturbances, anorexia, weight loss, malaise and headache [35]. In depression, we found that signs of the AP response, such as increased plasma Hp concentrations, were significantly and positively related to psychomotor retardation, anorexia, weight loss, anergy, loss of interest in work and activities and middle insomnia [36]. Thus, beside the possible role of proinflammatory cytokines in IRS activation and neuroendocrine changes in major depression, cytokines may also contribute to the development of specific depressive symptoms.

The IRS activation model accounts for the multicausal etiology of depression whereby external or psychosocial stressors (negative life events), and internal or organic stressors are considered to play a pivotal role in the etiology of depression

It is generally accepted that major depression has a multicausal etiology, whereby internal (organic factors) as well as external (psychosocial) stressors are involved. The high prevalence of depression in various medical conditions is reflected in the DSM-IV diagnostic category "depression due to a medical condition". Medical conditions, which are associated with immune activation, such as infections (e.g. herpes virus, HIV, Borna virus, influenza), autoimmune disorders and multiple sclerosis, show an increased incidence of depressive symptomatology or depression [37]. Other conditions which are also accompanied by an enhanced production of cytokines, e.g. stroke, Alzheimer's disease and the postpartum period often trigger depressive symptoms or major depression [37]. Thus, autoimmune-, perinatal-, infectious- and stroke-induced cytokine hyperproduction provides a reasonable mechanism to account for the significant higher rates of depression linked with these conditions. Based on the above findings, it is suggested that IRS activation is involved in the etiology and symptomatology of "depression due to a general medical condition" [37,38].

There is also evidence that external stressors may enhance the secretion of proinflammatory cytokines in experimental animals and humans. Psychological stressors, such as immobilization or open field stress, may increase the secretion of IL-6 in serum and IL-1 mRNA expression in the hypothalamus of the rodent [39–42]. Our laboratory found that academic examination stress in students significantly increased the stimulated production of various cytokines, such as IL-6, TNFα, IFNγ and IL-10 [43]. We found that there are two different profiles of cytokine secretion following psychological stress, i.e. a first with a predominant IFNγ versus IL-10 response (labeled IFNγ reactors) and a second with a predominant IL-10 versus IFNγ response (labeled IL-10 reactors) [44]. IFNγ reactors, but not IL-10 reactors, were characterized by a significant stress-induced increase in anxiety and depressive feelings and stimulated production of TNFα and IL-6 [44]. Thus, the predominant proinflammatory response to psychological stress is associated with significantly greater stress-induced feelings of anxiety and depression.

In conclusion, the IRS model of depression provides a mechanism to account for the organic (internal stress) as well as psychosocial (external stress) etiology of major depression, whereby both types of stressors cause depression, through IRS activation with increased secretion of proinflammatory cytokines, which interact with serotonergic turnover and HPA-axis activity.

References

1. Jefferson JW, Greist JH (1994) Mood disorders. In: Hales RE, Yudofsky SC, Talbott JA (eds) Textbook of psychiatry. The American Psychiatric Press, Washington, pp 470–476
2. Herbert TB, Cohen S (1993) Depression and immunity: a meta-analytic review. Psychol Bull 113:472–486
3. Maes M, Lambrechts J, Bosmans E, Jacobs J, Suy E, Vandervorst C, DeJonckheere C, Minner B, Raus J (1992) Evidence for a systemic immune

activation during depression: results of leukocyte enumeration by flow cytometry in conjunction with monoclonal antibody staining. Psychol Med 22:45–53

4. Muller N, Ackenheil M, Hofschuster E (1989) Altered T-cell number and reduced suppressor cell activity in patients with affective psychosis. In: Hadden JW, Masek K, Nistico G (eds) Interactions among central nervous system. Neuroendocrine and immune systems. Pythagora Press, Rome, pp 385–394
5. Sluzewska A, Rybakowski JK, Sobieska M, Bosmans E, Pollet H, Wiktorowicz K (1996) Increased levels of alpha-1-acid glycoprotein and interleukin-6 in refractory depression. Depression
6. Sluzewska A, Rybakowski J, Bosmans E, Sobieska M, Berghmans R, Maes M, Wiktorowicz K (1996) Indicators of immune activation in major depression. Psychiatr Res 64:161–167
7. Seidel A, Arolt V, Hunstiger M, Rink L, Behnisch A, Kirchner H (1996) Major depressive disorder is associated with elevated monocyte counts. Acta Psychiatr Scand 94:198–204
8. Maes M, Stevens W, Declerck L, Bridts C, Peeters D, Schotte C, Cosyns P (1993) A significantly increased expression of T cell activation markers in depression: additional evidence for an inflammatory process during that illness. Prog Psychopharmacolol Biol Psychiatry 17:214–255
9. Perini GI, Zara M, Carraro C, Tosin C, Gava F, Santucci MG, Valverde S, Defranchis G (1995) Psychoimmunoendocrine effects of panic disorder. Hum Psychopharmacol 10:461–465
10. Duch DS, Woolf JH, Nichol CA, Davidson JR, Garbutt JC (1984) Urinary excretion of biopterin and neopterin in psychiatric disorders. Psychiatr Res 11:83–89
11. Dunbar PR, Hill J, Neale TJ, Mellsop GW (1992) Neopterin measurement provides evidence of altered cell-mediated immunity in patients with depression, but not with schizophrenia. Psychol Med 22:1051–1057
12. Maes M, Scharpé S, Meltzer HY, Okayli G, Bosmans E, D'Hondt P, Vanden Bossche B, Cosyns P (1994) Increased neopterin and interferon γ secretion and lower availability of L-tryptophan in major depression: further evidence for activation of cell-mediated immunity. Psychiatr Res 54:143–160
13. Bonaccorso S, Lin A, Verkerk R, VanHunsel F, Libbrecht I, Scharpé S, DeClerck L, Stevens W, Biondi M, Janca A, Maes M (1998) Immune markers in fibromyalgia: comparison with major depressed patients and normal volunteers. J Affect Disord 48:75–82
14. Linnoila M, Whorton R, Rubinow DR, Cowdry RW, Ninan PT, Waters RN (1983) CSF prostaglandin levels in depressed and schizophrenic patients. Arch Gen Psychiatry 40:405–406
15. Calabrese JR, Skwerer RG, Barna B, Gulledge AD, Valenzuela R, Butkus A, Subichin S, Krupp NE (1986) Depression, immunocompetence, and prostaglandins of the E series. Psychiatr Res 17:41–47
16. Song C, Lin A, Bonaccorso S, Heide C, Verkerk R, Kenis G, Bosmans E, Scharpe S, Cosyns P, DeJong R, Maes M (1998) The inflammatory response system and the availability of tryptophan to the brain of patients with major depression and sleep disorders. J Affect Disord 49:211–219
17. Maes M, Meltzer HY, Bosmans E, Bergmans R, Vandoolaeghe E, Ranjan R, Desnyder R (1995) Increased plasma concentrations of interleukin-6,

soluble interleukin-6, soluble interleukin-2 and transferrin receptor in major depression. J Affect Disord 34:301–309
18. Seidel A, Arolt V, Hunstiger M, et al. (1995) Cytokine production and serum proteins in depression. Scand J Immunol 41:534–538
19. Seidel A, Arolt V, Hunstiger M, Rink L, Behnisch A, Kirchner H (1996) Increased CD56+ natural killer cells and related cytokines in major depression. Clin Immunol Immunopathol 78:83–85
20. Maes M, Wauters A, Verkerk R, Neels H, vanGastel A, Cosyns P, Scharpe S, Desnyder R (1996) Lower L-tryptophan availability in depression: a marker of a more generalized disorder in protein metabolism. Neuropsychopharmacol 15:243–251
21. Nassberger L, Traskman-Bendz L (1993) Increased soluble interleukin-2 receptor concentrations in suicide attempters. Acta Psychiatr Scand 88:48–52
22. Maes M (1993) Acute phase protein alterations in major depression: a review. Rev Neurosci 4:407–416
23. Joyce PR, Hawes CR, Mulder RT, Sellman JD, Wilson DA, Boswell DR (1992) Elevated levels of acute phase plasma proteins in major depression. Biol Psychiatry 32:1035–1041
24. Song C, Dinan T, Leonard BE (1994) Changes in immunoglobulin, complement and acute phase protein levels in the depressed patients and normal controls. J Affect Disord 30:283–288
25. Sluzewska A, Rybakowski JK, Sobieska M, Wiktorowicz K (1996) Concentrations and microheterogeneity glycophorms of alpha-1-acid glycoprotein in major depression. J Affect Disord 39:149–155
26. Sluzewska A, Nowakowska E, Gryska K, Mackiewicz A (1994) Haptoglobin levels in a chronic mild stress model of depression in rats before and after treatment. Eur Neuropharmacol P-1-18:302
27. Song C, Leonard BE (1994) An acute phase protein response in the olfactory bulbectomized rat: effect of sertraline treatment. Med Sci Res 22:313–314
28. Sluzewska A, Rybakowski JK, Laciak M, Mackiewicz A, Sobieska M, Wiktorowiz K (1995) Interleukin-6 serum levels in depressed patients before and after treatment with fluoxetine. Ann NY Acad Sci 762:474–476
29. Maes M, Delange J, Ranjan R, Meltzer HY, Desnyder R, Cooremans W, Scharpé S (1997) Acute phase proteins in schizophrenia, mania and major depression: modulation by psychotropic drugs. Psychiatr Res 66:1–11
30. Xia Z, DePierre JW, Nassberger L (1996) Tricyclic antidepressants inhibit IL-6, IL-1β and TNF-α release in human blood monocytes and IL-2 and interferon-γ in T cells. Immunopharmacol 34:27–37
31. Maes M, Bosmans E, Meltzer HY, Scharpé S, Suy E (1993) Interleukin-1ß: a putative mediator of HPA-axis hyperactivity in major depression? Am J Psychiatry 150:1189–1193
32. Pariante CM, Pearce BD, Pisell TL, et al. (1997) Steroid-independent modulation of the glucocorticoid receptor: relevance for psychiatry and immunology. In: Abstract Book WPA Symposium "Psychiatry in the Three Ages of Man". WPA, Rome, pp 21
33. Kent S, Bluthé R-M, Kelley KW, Dantzer R (1992) Sickness behavior as a new target for drug development. TiPS 13:24–28

34. Yirmiya R (1996) Endotoxin produces a depressive-like episode in rats. Brain Res 711:163–174
35. McDonald EM, Mann AH, Thomas HC (1987) Interferons as mediators of psychiatric morbidity. Lancet 2:1175–1178
36. Maes M, Meltzer HY, Scharpé S, Cooreman W, Uyttenbroeck W, Suy E, Vandervorst C, Raus J, Cosyns P (1993) Psychomotor retardation, anorexia, weight loss, sleep disturbances and loss of energy: psychopathological correlates of hyperhaptoglobinemia in major depression. Psychiatr Res 47:229–241
37. Maes M (1997) The immune pathophysiology of major depression. In: Honig A, van Praag HM (eds) Depression: neurobiological, psychopathological and therapeutic advances. John Wiley, London, pp 197–215
38. Yirmiya R (1997) Behavioral and psychological effects of immune activation: implications for depression due to a general medical condition. Curr Opin Psychiatry 10:470–476
39. LeMay LG, Vander AJ, Kluger MJ (1990) The effects of psychological stress on plasma interleukin-6 activity in rats. Physiol Beh 47:957–961
40. Zhou D, Kusnecov AW, Shurin MR, De Paoli M, Rabin B (1993) Exposure to physical and psychological stressors elevates plasma interleukin-6: relationship to the activation of hypothalamic-pituitary-adrenal axis. Endocrinol 133:2523–2530
41. Soszynski D, Kozak W, Conn CA, Rudolph K, Kluger MJ (1996) Beta-adrenoceptor antagonists suppress elevation in body temperature and increase plasma IL-6 in rats exposed to open field. Neuroendocrinol 63:459–467
42. Minami M, Kuraishi Y, Yamaguchi T, Nakai S, Hirai Y, Satoh M (1991) Immobilization stress induces interleukin-1 beta mRNA in the rat hypothalamus. Neurosci Lett 25:254–256
43. Maes M, Song C, Lin A, Gabriels L, DeJongh R, Van Gastel A, Kenis G, Bosmans E, DeMeester I, Benoyt I, Neels H, Demedts P, Janca A, Scharpe S, Smith RS (1998) The effects of psychological stress on humans: increased production of proinflammatory cytokines and a Th-1-like response in stress-induced anxiety. Cytokine 10:313–318
44. Maes M, Song C, Lin A, DeJong R, Kenis G, Bosmans E, DeMeester I, Neels H, Scharpe S (1998) Immune and clinical correlates of psychological stress-induced production of interferon-γ and IL-10 in humans. In: Plotnikoff NP et al. (eds) Cytokines, stress, and immunity. CRC Press, Boca Raton, pp 39–50

Correspondence: Prof.Dr.M.Maes, M.D., Ph.D., Clinical Research Center for Mental Health, University Department of Psychiatry, AZ Stuivenberg, 267 Lange Beeldekensstraat, 2060 Antwerp, Belgium

Methodological concerns in the study of the immune system in schizophrenia

D.P. van Kammen[1,2], **C.G. McAllister-Sistilli**[1,2,3], **M.E. Kelley**[1], **J.A. Gurklis**[1,2], and **J.K. Yao**[1,2]

[1] VA Pittsburgh Healthcare System, and Departments of [2] Psychiatry and [3] Pathology, University of Pittsburgh School of Medicine, Pittsburgh, PA, USA

Summary

Background. Studies of immune measures in schizophrenia have shown differences with control subjects on a number of measures. However, immune measures are often not detectable in the serum or CSF of human samples and in many studies it is not clear how this was dealt with in the analysis. *Methods.* CSF IL-6 was measured by ELISA in 61 drug free male schizophrenic (DSM-IIIR) patients and 25 well-screened healthy male control subjects. Serum IL-6 was measured in 43 of the 61 patients, and in 16 control subjects. Data were analyzed with and without non-detectable values to determine if this affected the results. *Results.* Using a value of "0" for non-detectable values resulted in significant differences between patients and controls in plasma interleukin-6. However, if these values were treated as missing, the difference was not significant. *Conclusions.* The treatment of non-detectable immune measures has demonstrable effects on the analysis and interpretation of the results.

Introduction

Immune studies in schizophrenia are frequently plagued by assay issues which have not yet been addressed consistently. In our study of interleukin-2 and interleukin-1α in relapse prediction in schizophrenia (McAllister et al., 1995), variance issues were raised that made the use of parametric tests tenable for these measures of biochemistry. In uncovering the reasons for the variance differences, however, it was determined that the distribution of non-detectable values was the source of the discrepancy and that using those values as "low" rather than missing data, could have effects on the analysis. For this finding, both treatments of the data resulted in the same conclusions.

Elevated levels of serum IL-6, or the frequency of individuals with such levels, have been reported in schizophrenia, however the treatment of non-detectable values has not been consistently reported across studies. The majority of studies report the detection limits of the assay, but do not indicate if any samples were below this level and what was done with them (Maes et al., 1995a,b, 1997; Pollmacher et al., 1996; Shintani et al., 1991; Barak et al., 1995; Baker et al., 1996). Presumably values that were not detectable were excluded from these analyses. Of the studies that have acknowledged these non-detectable values, Katila et al. (1994a,b) reported no detectable values in schizophrenic patients, while Ganguli et al. (1994) reported 19% non-detectable values in patients and Naudin et al. (1996) reported 73% non-detectable values in patients. Attempts to incorporate information from these assays have included strategies of using a "0" value for the non-detectable levels (Naudin et al., 1997; McAllister et al., 1995), and "spiking" the assay with IL-6 to raise the baseline level for detection (Frommberger et al., 1997).

Thus, clearly there are a significant number of both schizophrenic and control subjects who have no detectable levels of interleukins, yet this has not yet been addressed as an issue in these studies. If the ELISA is valid, non-detectable values should be considered "low" rather than missing or invalid data. Moreover, if the percentage of non-detectable values is significantly different in the patient and control samples, this has serious implications for the results of the analysis. In Ganguli et al. (1994) the percentages were 19% and 23% respectively, and thus it was concluded that removing these data would not alter the analyses. However, most studies fail to report the incidence of non-detectable values, thus it can not be ensured that the subsequent analyses are valid.

As an example we analyzed our IL-6 data in patients and controls using both strategies for the non-detectable values. Specifically, we repeated the analyses with the non-detectable values treated as levels of "0" and as missing data.

Methods

Schizophrenic patients

Sixty-one (61) physically healthy, male psychiatric patients with a diagnosis of chronic schizophrenia or schizoaffective disorder (DSM-IIIR) (American Psychiatric Association, 1987), on antipsychotic maintenance treatment, participated in the study following admission to the Schizophrenia Research Unit at the VA Pittsburgh Healthcare System in Pittsburgh, PA, and giving informed consent. The staff reviewed with each patient – and relevant others – prodromes and previous relapses, the risk, and what actions would be taken. Fifty-six subjects were European-American, four were African-American, and one was Latin American. All subjects were screened with a complete physical, neurologic, and psychiatric evaluation conducted by board certified psychiatrists. Trained staff obtained the diagnostic data from a structured interview using the SADS-L (Spitzer and Endicott, 1979) or the SCID (Spitzer et al., 1989). Patients who met DSM-IIIR criteria for alcohol or substance abuse/dependence were excluded at the time of the study, unless they had been in remission for at least 6 months. Schizophrenia diagnoses included paranoid (N = 30), undifferentiated (N = 22), disorganized (N = 2), and schizoaffective disorder, mainly schizophrenic (N = 7).

Design and evaluation

Upon admission, the subject's medication was gradually converted to haloperidol if they were not already being treated with haloperidol. Within 6 ± 4 days of procedures on medication, haloperidol was replaced overnight with placebo in unmarked capsules. Drug-free procedures took place after 38 ± 13 days (range: 14–67 days). A three day mean increase in 3 points over the baseline (pre-drug withdrawal) psychosis ratings done by the nursing staff determined relapse during the drug free period as previously validated (van Kammen et al., 1989, 1994). Of the 61 subjects, 31 met relapse criteria within six weeks.

Normal control subjects

Subjects were recruited from the community through newspaper ads. They were extensively screened to exclude histories (including family history) of psychiatric, immunologic, major physical disorders and recent substance abuse. They received a diagnostic and psychiatric interview, a similar workup as our patients, and a drug screen prior to testing. Eleven (11) of the 25 subjects had both CSF and plasma measures taken; an additional 4 subjects received plasma draws only.

Lumbar puncture (LP) and blood draws

Spinal fluid was obtained according to established procedures (van Kammen and Sternberg, 1980) with patients in the lateral decubitus position between 7:30 and 8:30 a.m. after bed rest and fasting since 10:30 pm the previous night. CSF was collected on ice in a 12 ml pool, that was divided into 0.5 and 1.0 ml aliquots. Blood for plasma IL-6 was drawn during the week of the LP. Samples were stored at –80°C until being assayed.

Assays

CSF and plasma IL-6 levels were measured by ELISA using a commercially available kit (CYTImmune, Sciences, Inc., College Park, MD). The sensitivity is

Table 1. Test of independence: percentage of non-detectable values

Variable	Schizophrenics	Control subjects	χ^2	p
CSF IL-6	27/61 (44%)	15/25 (60%)	1.76	Ns
Plasma IL-6	11/43 (26%)	11/16 (69%)	9.29	0.002

0.195 ng/ml, with intraassay and interassay coefficients of 7.0 % and 11.3% respectively. In addition to a standard curve ranging from 0.195 to 200 ng/ml, a reference standard for IL-6 (provided by the Biological Response Modifiers Program, National Cancer Institute) was used in evaluating the assay. Storage time was not associated with CSF or serum IL-6 concentrations.

Statistics

The primary concern in the statistical analysis of the IL-6 data was the treatment of non-detectable values determined by our assay. Because the number of non-detectable values was not negligible, our analyses included tests of independence (chi-square) of the distribution of non-detectable values as well as nonparametric tests for comparison between groups (Mann-Whitney) using the value of 0 for non-detectable.

Results

Table 1 shows the tests of significance of the percentages of non-detectable values. It can be seen that for CSF IL-6, these percentages were not significantly different and thus analysis of the data should be valid using either method. However, this is not the case for plasma IL-6. The control subjects have a significantly higher percentage of non-detectable values and thus treatment of these values has profound effects on the results. If the non-detectable values are truly "low" then this would indicate that schizophrenic patients most likely have higher plasma IL-6 values overall. In Table 2, the results of the two approaches to analysis are presented. It is clear that the results of the plasma analysis are significant if the non-detectable values are used, while they are not if the non-detectable values are considered missing.

Discussion

Our finding of an elevation in plasma levels of IL-6 is consistent with the majority of studies of schizophrenia (Maes et al., 1995a,b, 1997; Pollmacher et al., 1996; Ganguli et al., 1994; Shintani et al., 1991). However, this result rests on the treatment of values not detected by

Table 2. Interleukin-6 (ng/ml): schizophrenic patients compared to control subjects with and without non-detectable levels. Sample sizes shown in the denominator

With non-detectable values set to 0:			Variance test		Wilcoxon test	
Variable	Schizophrenics	Control subjects	F	p	Z	p
CSF IL-6	0.53 (0.70)	0.25 (0.43)	2.62	0.011	−1.77	0.077
Plasma IL-6	3.25 (3.16)	1.20 (1.96)	2.60	0.048	−2.74	0.006
With non-detectable values as missing:			**Variance test**		**Wilcoxon test**	
Variable	Schizophrenics	Control subjects	F	p	Z	p
CSF IL-6	0.95 (0.70)	0.62 (0.50)	1.95	ns	−1.54	ns
Plasma IL-6	4.37 (2.91)	3.84 (1.30)	4.97	ns	−0.36	ns

the assay. While a reasonable percentage of our values were non-detectable (Table 1), we nevertheless observed meaningful values in most patients. Our approach in this study was to accept undetectable values as being "low" but to make no assumptions about the level assigned to them by using non-parametric procedures. It is not clear that this is the best approach; undetectable levels may not be low but undeterminable and in that case, should be excluded. Some investigators have excluded patients whose values fell below the sensitivity of their assay, which may skew the results if the allocations are different between patients and control subjects. Thus, more studies need to report this "missing" data before the best approach to analysis can be determined.

IL-6 measurements can be influenced by the presence of serum proteins that form complexes with IL-6, at least with conventional assays (May et al., 1992; Ledur et al., 1995), thus not all studies will find detectable levels. In addition, Ndubuisi et al. (1998) have conclusively shown that the distinct classes of IL-6, differentiated based on molecular weight, have a unique detection pattern dependent on the ELISA or bioassay used. These authors also conclude that binding molecules in the blood, including autoantibodies and soluble IL-6 receptors will influence IL-6 determinations. The IL-6 assay chosen for our study measures total IL-6, both free and complexed, so that the potential influence of binding factors on our serum measurements is negligible. We conclude therefore, that the non-detectable levels are indeed low levels. The presence of these IL-6 binding factors in CSF has not been examined.

It is highly possible that these technical issues play a role in the discrepancies in the plasma and CSF findings of IL-6 in schizophrenia. Clearly, the handling of these non-detectable values in statistical analyses is of issue and will affect the interpretation of results. Future studies in this area should make note of this methodological issue and show how the data were used.

References

American Psychiatric Association (1987) Diagnostic and statistical manual of mental disorders, 3rd edn, rev. American Psychiatric Press, Washington, DC

Baker I, Masserano J, Wyatt RJ (1996) Serum cytokine concentration in patients with schizophrenia. Schizophr Res 20:199–203

Barak V, Barak Y, Levine J, Nisman B, Roisman I (1995) Changes in Interleukin-1β and soluble interleukin-2 receptor levels in CSF and serum of schizophrenic patients. J Basic Clin Physiol Pharmacol 6:61–69

Frommberger UH, Bauer J, Haselbauer P, Fraulin A, Riemann D, Berger M (1997) Interleukin-6-(IL-6) plasma levels in depression and schizophrenia: comparison between the acute state and after remission. Eur Arch Psychiat Clin Neurosci 247:228–233

Ganguli R, Yang Z, Shurin G, Chengappa RKN, Brar JS, Gubbi AV, Rabin BS (1994) Serum interleukin-6 concentration in schizophrenia: elevation associated with duration of illness. Psychiatry Res 51:1–10

Katila H, Hurme M, Wahlbeck K, Appelberg B, Rimon R (1994a) Plasma and cerebrospinal fluid interleukin-1β and interleukin-6 in hospitalized schizophrenic patients. Neuropsychobiol 30:20–23

Katila H, Appelberg B, Hurme M, Rimon R (1994b) Plasma levels of interleukin-1β and interleukin-6 in schizophrenia, other psychoses, and affective disorders. Schizophr Res 12:29–34

Ledur A, Fitting C, David B, Hamberger C, Cavaillon J-M (1995) Variable estimates of cytokine levels produced by commercial ELISA kits: results using international cytokine standards. J Immunol Meth 186:171–179

Maes M, Meltzer HY, Buckley P, Bosmans E (1995a) Plasma-soluble interleukin-2 and transferrin receptor in schizophrenia and major depression. Eur Arch Psychiatry Clin Neurosci 244:325–329

Maes M, Bosmans E, Calabrese J, Smith R, Meltzer H (1995b) Interleukin-2 and interleukin-6 in schizophrenia and mania: effects of neuroleptics and mood stablizers. J Psychiatr Res 29:141–152

Maes M, Bosmans E, Kenis G, De Jong R, Smith RS, Meltzer HY (1997) In vivo immunomodulatory effects of clozapine in schizophrenia. Schizophr Res 26:221–225

May LT, Viguet H, Kenney JS, Ida N, Allison AC, Sehgal PB (1992) High levels of "complexed" interleukin-6 in human blood. J Biol Chem 267:19698–19704

McAllister CG, van Kammen DP, Rehn TJ, Miller AL, Gurklis J, Kelley ME, Yao J, Peters JL (1995) Increases in CSF levels of interleukin-2 in schizophrenia: effects of recurrence of psychosis and medication status. Am J Psychiatry 159:1291–1297

Naudin J, Mege JL, Azorin JM, Dassa D (1996) Elevated circulating levels of IL-6 in schizophrenia. Schizophr Res 20:269–273

Naudin J, Kapo C, Giusano B, Mege JL, Azorin JM (1997) A differential role for interleukin-6 and tumor necrosis factor-alpha in schizophrenia? Schizophr Res 26:227–233

Ndubuisi MI, Patel K, Rayanade RJ, Mittelman A, May LT, Seghal PB (1998) Distinct classes of chaperoned IL-6 in human blood: differential immunological and biological availability. J Immunol 160:494–501

Pollmacher T, Hinze-Selch D, Mullington J (1996) Effects of clozapine on plasma cytokine and soluble cytokine receptor levels. J Clin Psychopharmacol 16:403–409

Shintani F, Shigenobu K, Maruo N, Nakaki T, Nibuya M, Suzuki E, Kinoshita N, Yagi G (1991) Serum interleukin-6 in schizophrenic patients. Life Sci 49:661–664

Spitzer RL, Endicott J (1979) The schedule for affective disorders and schizophrenia-lifetime version, 3rd ed. New York State Psychiatric Institute, Biometrics Research, New York

Spitzer RL, Williams JBW, Gibbon M, First MB (1989) Structured clinical interview for DSM-III-R (SCID). New York State Psychiatric Institute, New York

van Kammen DP, Sternberg DE (1980) CSF studies in schizophrenia. In: Wood JH (ed) Neurobiology of cerbrospinal fluid, vol I. Plenum Press, New York, pp 719–742

van Kammen DP, Peters JL, van Kammen WB, Nugent A, Goetz KL, Yao J, Linnoila M (1989) CSF norepinephrine in schizophrenia is elevated prior to relapse after haloperidol withdrawal. Biol Psychiatry 26:176–188

van Kammen DP, Ågren H, Yao JK, O'Connor DT, Gurklis JA, Peters JL (1994) Noradrenergic activity and prediction of psychotic relapse following haloperidol withdrawal in schizophrenia. Am J Psychiatry 151:379–384

Correspondence: Dr. D. P. van Kammen, Global Clinical Research & Development, The R. W. Johnson Pharmaceutical Research Institute (PRI), 920 Route 202, Room 2312, P. O. Box 300, Raritan, NJ 08869-0602, USA

Psychoimmunological alterations in schizophrenia: how do confounding variables influence the findings?

M. Rothermundt, V. Arolt, M. Peters, and **H. Kirchner**

Department of Psychiatry and Institute of Immunology and Transfusion Medicine, School of Medicine, Universities of Münster and Lübeck, Federal Republic of Germany

Summary

Several groups have contributed to the progress of immunological research in schizophrenia. Early studies demonstrated partly divergent results. The recently improved control concerning confounding variables has lead to increasingly consistent findings. The influence of psychotropic medication, serum cortisol levels, different immunological techniques, sex differences and psychopathology on psychoimmunological findings in schizophrenia is discussed.

Novel findings due to technical advances have strengthened the hypothesis that immunological dysfunction may contribute towards the multifactoral etiology of schizophrenia (Kirch, 1993; Syvälahti, 1994; Wright et al., 1993). Since the early work of Bruce and Peebles (1904) who reported increased leukocyte counts, especially during the acute phase of illness, numerous studies have focused on the role of cellular immunity in schizophrenia. One particular line of research has focused on investigating the numbers of immunocompetent cells. Nyland et al. (1980) showed a decreased number of T-lymphocytes in acute but not in chronic schizophrenics. An elevated number of T-cells was reported by two groups (Henneberg et al., 1990; Ackenheil et al., 1991; Müller et al., 1991, 1993) while others (including ours) demonstrated no changes in T-cell numbers (Villemain et al., 1989; Masserini et al., 1990; Achiron et al., 1994; Sasaki et al., 1994; Hornberg et al., 1995; Wilke et al., 1996). Inconsistent results have also been reported concerning several T-lymphocyte subtypes, B-lymphocytes and natural killer cells (NK-cells).

Another approach has focused on cytokine production. Cytokines are produced by leukocytes; they regulate the differentiation and activation of immunologically active cells and control the communication between the immunocompetent cells. The production of cytokines after mitogen stimulation can be regarded as a functional marker of cellular immunity.

Research on cytokines in schizophrenia has yielded increasingly consistent results. In response to mitogen stimulation, Kolyaskina (1983) showed a reduced proliferative activity of T-lymphocytes while serum levels of interleukin-2 (IL-2) were found to be unchanged (Barak et al., 1995; Gattaz et al., 1992). Furthermore, several groups (including our own) have reported that the production of IL-2 by lymphocytes after mitogen stimulation is significantly lower in schizophrenics than it is in non-psychotic controls (Villemain et al., 1989; Hornberg et al., 1995; Bessler et al., 1995; Ganguli and Rabin, 1989; Ganguli et al., 1992,1995; Rothermundt et al., 1996). The decreased IL-2 production after mitogen stimulation of acutely ill schizophrenics compared to patients in remission suggests that IL-2 is a state rather than a trait marker for the acute illness (Ganguli and Rabin, 1993; McAllister et al., 1995; Arolt et al., 1997). In a functional sense, interferon gamma (IFN-γ) and IL-2 are closely related since the production of IFN-γ is stimulated by IL-2. Despite this relationship the results concerning IFN-γ in schizophrenia have so far been less impressive than those regarding IL-2. Serum levels of IFN-γ are not on the whole elevated (Gattaz et al., 1992; Becker et al., 1990; Schindler et al., 1986) although higher levels were reported by Preble and Torrey (1985). A tendency towards decreased production after stimulation has been reported on several occasions (Hornberg et al., 1995; Katila et al., 1989; Moises et al., 1985). In a group of psychotic patients which included schizophrenics and depressed patients, Inglot et al. (1994) showed a decreased production of IFN-γ.

For a correct interpretation of the at least partially inconsistent immunological findings in schizophrenia it remains an important task to distinguish the influence of the following interfering variables that might influence the results and complicate interpretations:

1. Cortisol levels
2. Medication
3. Immunological technique
4. Sex differences
5. Psychopathology

1. Cortisol levels

There are complex interactions between the different hormones, especially cortisol, and the cellular as well as humoral immunological subsystems. Cortisol inhibits inflammation, decreases the number of lymphocytes in the peripheral blood, prevents leukocytes from migrating into the tissue and decreases the production of antibodies and cytokines. Only in a few psychoimmunological studies on schizophrenic patients has cortisol been measured. While investigating both schizophrenics and a subgroup of schizophrenics with signs of autoimmunity Ganguli and Rabin (1993) showed that serum cortisol was significantly higher in patients with low IL-2 production. However, a negative correlation between cortisol and IL-2 production was demonstrated only in autoantibody positive patients. Sasaki et al. (1994) showed normal cortisol levels on hospital admission but increased levels after 8 weeks of treatment. In a study investigating 44 acute schizophrenic patients all patients displayed normal serum cortisol levels (Rothermundt et al., 1998); no differences between patients and controls were detected, and there were no correlations between cortisol levels and either IL-2 or IFN-γ production.

Measurements of cortisol levels after applying the dexamethasone inhibition test could further our knowledge concerning a potential role for cortisol in bringing about the observed changes in cytokine production.

2. Medication

The influence of neuroleptic medication on immunological processes remains unclear at the present point in time. Much psychoimmunological information has been obtained from both medicated and non-medicated schizophrenic patients. However, certain neuroleptic drugs seem to influence the immune system. Typical antipsychotic drugs (e.g. haloperidol) seem to have negative immunoregulatory effects while atypical neuroleptics (e.g. clozapine and risperidone) showed complex in vivo immunomodulatory effects (Maes et al., 1994, 1995, 1996; Pollmächer et al., 1995, 1996; Smith and Maes, 1995; Tarazona et al., 1995).

No differences in the numbers of T-helper cells could be shown in either unmedicated (Villemain et al., 1989; Masserini et al., 1990; Achiron et al., 1994; Nikkilä et al., 1995) or medicated schizophrenic patients (Masserini et al., 1990; Sasaki et al., 1994; Hornberg et al., 1995; Wilke et al., 1996) relative to healthy individuals. Masserini et al. (1990) demonstrated normal CD-8 positive cell numbers in medicated schizophrenic patients but higher than normal numbers in drug-free patients while Ackenheil et al. (1991) observed higher numbers in medicated schizophrenics. Monteleone et al. (1991), Müller et al. (1991) and Rapaport et al. (1991) showed no differences between medicated and unmedicated patients concerning the mitogen response.

B-lymphocytes in medicated patients were reported to be either unchanged (Masserini et al., 1990; Ganguli and Rabin, 1993) or elevated compared to healthy controls (McAllister et al. (1989) in 30% of the patients), (Rogozhnikova et al., 1993). In unmedicated schizophrenics, both unchanged (Ganguli and Rabin, 1993) and elevated (Masserini et al., 1990) numbers were found.

A decreased IL-2 production has been shown in medicated (Hornberg et al., 1995; Ganguli et al., 1992) and non-medicated (Villemain et al., 1989; Bessler et al., 1995; Ganguli et al., 1995) schizophrenics.

In several studies Rapaport et al. showed an increased serum level of the sIL-2R in medicated (Rapaport et al., 1989, 1993, 1994a, b) as well as unmedicated (Rapaport et al., 1994a, b) schizophrenic patients. Ganguli's group confirmed these results (1989, 1993). Maes et al. (1994, 1995) demonstrated increased sIL-2R levels in both unmedicated and medicated patients suffering from schizophrenia, but clozapine treatment seemed to increase such levels (Maes et al., 1994). This was confirmed by Pollmächer et al. (1995, 1996) who showed an increase of sIL-2R in patients treated with clozapine but unchanged levels in schizophrenics treated with haloperidol.

In order to find out how unknown serum factors such as medication influence the cytokine production in vitro we applied a "criss-cross" technique: Cells and serum of patients and controls were separated before the cells underwent a washing procedure. The cells of each patient were incubated with the serum of the matched control and vice versa. In this way the possibility that the cells of control patients could be influenced by the sera of schizophrenic patients could be investigated.

Cytokine production by patient cells incubated in control serum did not differ from that of patient cells incubated in their own serum after the separation procedure. Likewise, there was no difference in cytokine production after mitogen stimulation of control cells by patient serum as compared to control cells cultured in their own serum after the washing procedure. By applying the criss-cross technique it could be shown that serum factors did not influence cytokine production in vitro. The leukocytes of patients could not be positively influenced by adding healthy control serum and the control cells were unaffected by the patient serum (Rothermundt et al., 1998).

With this in vitro technique we were not able to investigate the in vivo influences of serum factors such as medication or hormones. However, evidence is provided that there is a genuine immunological dysfunction in schizophrenia since the changes in cytokine production are not likely to be caused by unknown serum factors.

3. Immunological technique

Another difficulty in interpreting psychiatric patient immunological findings, especially those concerning cytokine production, arises from differences in the various immunological techniques applied. Some authors reported results from whole blood assays (Hornberg et al., 1995; Wilke et al., 1996; Ganguli et al., 1992, 1993, 1995; Rothermundt et al., 1996; Moises et al., 1985; Inglot et al., 1994) while others published results from cells which had been separated by a washing procedure (Villemain et al., 1989; Bessler et al., 1995). Dealing with this aspect we found that the separation procedure for cells and serum resulted in a decreased cytokine production compared to that measured in the whole blood (Rothermundt et al., 1998). Incubating patient or control cells that had undergone the washing procedure in their own serum also led to a lower leukocyte cytokine production than incubation in a whole blood assay. Such differences in mitogen stimulated leukocyte cytokine production, depending on whether cells were cultured as whole blood in situ or with autologous serum after fractionation, were demonstrated for both normal and schizophrenic individuals (Rothermundt et al., 1998). The true state of immunological alterations in psychiatric disorders would be clarified if future studies by different groups would apply comparable immunological techniques. We favor the whole blood assay described by Kirchner et al. (1982) as do most other groups. In this technique the cells remain in their natural surroundings and have access to the serum factors they need for a sufficient cytokine production. Furthermore, the cells do not undergo a washing procedure which apparently stresses the cells since the separated cells produce only 10% of the cytokines which can be produced by cells cultured in whole blood. Higher amounts of cytokines are easier to detect by an ELISA technique.

4. Sex differences

Another variable that could confound the reported findings of immunological alterations in psychiatric disorders is the sex of the investigated subjects. A gender difference in the production of IFN-γ has previously been demonstrated by Kita et al. (1991) in healthy subjects. We confirmed these findings for healthy as well as schizophrenic subjects. Female schizophrenic patients as well as female controls showed lower levels of IFN-γ production after stimulation than their males counterparts ($p <= 0.01$) (Rothermundt et al., 1998).

In our recent study there was a tendency towards lower sIL-2R levels in females as compared to males although this difference was not significant. Moreover, female patients showed lower levels than their matched controls while male patients demonstrated higher levels compared to their matched controls (Rothermundt et al., 1998). In three studies which reported increased sIL-2R levels and in which the sex distribution was indicated the majority of the subjects studied were male (Rapaport et al., 1994b; Maes et al., 1994, 1995). Only one study reported an equally balanced sex distribution and increased sIL-2R levels in schizophrenics (Rapaport et al., 1994a).

It is therefore important that reports on psychoimmunological findings should always specify the sex distribution and show the results obtained from male and female patients separately.

5. Psychopathology

Previous studies have shown that immunological findings in schizophrenia differ according to the state of illness. For example acutely ill schizophrenics showed a decreased IFN-γ production (Wilke et al., 1996; Rothermundt et al., 1996; Arolt et al., 1997) while schizophrenics in remission showed no difference in IFN-γ production after stimulation (Wilke et al., 1996). Many of the reported investigations on cytokines and the numbers of immunocompetent cells differentiated between acute and remitted schizophrenics (Becker et al., 1990; Villemain et al., 1989; Ganguli et al., 1992, 1995; Katila et al., 1989; Barak et al., 1995; Maes et al., 1994) while in other studies patients were not differentiated according to the state of their illness (e.g. Bessler et al., 1995; Moises et al., 1985; Achiron et al., 1994; McDaniel et al., 1992; Ghosh and Chattopadhyay, 1993; McAllister et al., 1989; Rapaport et al., 1991). A few studies included only first episode schizophrenic patients (Becker et al., 1990; Ganguli et al., 1995; Maes et al., 1994). In some studies, psychotic patients with schizophrenic, schizoaffective and affective disorders were all compiled into one sample (Inglot et al., 1994; Preble and Torrey, 1985; Katila, 1994; Maes et al., 1995; Ackenheil et al., 1991; Müller et al., 1991, 1993; Masserini et al., 1990; Ganguli and Rabin, 1993; Nikkilä et al., 1995; Abdeljaber et al., 1994).

Indeed, most studies included different schizophrenic subtypes in their sample (Villemain et al., 1989; Maes et al., 1995; Katila et al., 1989; Bessler et al., 1995; Ganguli et al., 1992; Barak et al., 1995; Rothermundt et al., 1996; Hornberg et al., 1995) while others restricted their analysis to patients with paranoid schizophrenia (Gattaz et al., 1992) or specifically indicated results for the different subtypes of this disease grouping (Wilke et al., 1996).

In a study on 44 schizophrenic patients suffering from an acute psychotic episode we divided the patients into two subgroups. One subgroup consisted of patients suffering from the paranoid subtype of schizophrenia, while the other consisted of patients with disorganized schizophrenia as defined by the DSM IV protocol. Cytokine production by stimulated immunocompetent cells did not differ significantly between the two subgroups (Table 1).

Only a few studies were able to demonstrate correlations between psychopathological or clinical features and immunological findings.

Table 1. Cytokines and sIL-2R in pg/ml in different subtypes of schizophrenia

Schizophrenic subtype	IL-2	IL-2 controls	IFN-γ	IFN-γ controls	sIL-2R	sIL-2R controls
Paranoid subtype	121.8	277.8	2007.9	5569.8	2010.7	2143.6
Disorganized subtype	129.5	224.0	2146.6	5899.4	1769.0	1710.9
p	n.s.	n.s.	n.s.	n.s.	n.s.	n.s.

IL-2 Interleukin-2, *IFN-γ* interferon gamma, *sIL-2R* soluble Interleukin-2 receptor, *n.s.* not significant

Ganguli et al. (1995) reported a positive correlation between IL-2 production and age of onset, and a negative correlation between IL-2 production and negative symptom scores. In the study of Rapaport et al. (1993), the increased serum level of the sIL-2R was associated with both early onset of disease and negative symptoms. In our study we were not able to reproduce these findings.

Conclusion

In order to improve the comparability of future studies on immunological alterations in schizophrenia it is important to specifically indicate the immunological and psychiatric methods used. Variables which might interfere with the immunological data such as medication status, cortisol levels, sex differences, schizophrenic subtypes, course of illness or state of illness should be taken into consideration. Although today there is little doubt that immunological alterations are present in schizophrenia, the quality of analysis is dependent on controlling for the relevant confounders.

References

Abdeljaber MH, Nair MPN, Schork MA, Schwartz SA (1994) Depressed natural killer cell activity in schizophrenic patients. Immunol Invest 23:259–268

Achiron A, Noy S, Pras E, Lereya J, Hermesh H, Laor N (1994) T-cell subsets in acute psychotic schizophrenic patients. Biol Psychiatry 35:27–31

Ackenheil M, Hofschuster E, Müller N (1991) Cellular immunity and immune genetics in schizophrenia. Biol Psychiatry 2:577–580

Arolt V, Weitzsch C, Wilke I, Nolte A, Pinnow M, Kirchner H (1997) Production of interferon-gamma in families with multiple occurence of schizophrenia. Psychiatry Res 66:145–151

Barak V, Barak Y, Levine J, Nisman B, Roisman I (1995) Changes in interleukin-1 beta and soluble interleukin-2 receptor levels in CSF and serum of schizophrenic patients. J Basic Clin Physiol Pharmacol 6:61–69
Becker D, Kritschmann E, Floru S, Shlomo-David Y, Gotlieb-Stematsky T (1990) Serum interferon in first psychotic attack. Br J Psychiatry 157:136–138
Bessler H, Levental Z, Karp L, Modai I, Djaldetti M, Weizman A (1995) Cytokine production in drug-free and neuroleptic-treated schizophrenic patients. Biol Psychiatry 38:297–302
Bruce LC, Peebles AMS (1904) Quantitative and qualitative leukocyte counts in various forms of mental disease. J Ment Sci 50:409–417
Ganguli R, Rabin BS (1989) Increased serum interleukin-2 receptor concentration in schizophrenic and brain-damaged subjects. Arch Gen Psychiatry 46:292
Ganguli R, Rabin BS (1993) CD5 positive lymphocytes in schizophrenia: no alteration in numbers or percentage as compared with control subjects. Psychiatry Res 48:69–78
Ganguli R, Brar JS, Solomon W, Chengappa KN, Rabin BS (1992) Altered interleukin-2 production in schizophrenia: association between clinical state and autoantibody production. Psychiatry Res 44:113–123
Ganguli R, Jaspreet SB, Chengappa KNR, DeLeo M, Yang ZW, Shurin G, Rabin BR (1995) Mitogen-stimulated interleukin-2 production in never-medicated first episode schizophrenic patients. Arch Gen Psychiatry 52:668–672
Gattaz WF, Dalgalarrondo P, Schröder HC (1992) Abnormalities in serum concentrations of interleukin-2, interferon-α and interferon-γ in schizophrenia not detected. Schizophr Res 6:237–241
Ghosh N, Chattopadhyay U (1993) Enhancement of immune response by phenothiazine administration in vivo. In Vivo 7:435–440
Henneberg A, Riedl B, Dumke HO, Kornhuber HH (1990) T-lymphocyte subpopulations in schizophrenic patients. Eur Arch Psychiatr Neurol Sci 239:283–284
Hornberg M, Arolt V, Wilke I, Kruse A, Kirchner H (1995) Production of interferons and lymphokines in leukocyte cultures of patients with schizophrenia. Schizophr Res 15:237–242
Inglot AD, Leszek J, Piasecki E, Sypula A (1994) Interferon responses in schizophrenia and major depressive disorders. Biol Psychiatry 35:464–473
Katila H, Cantell K, Hirvonen S, Rimón R (1989) Production of interferon-α and -γ by leukocytes from patients with schizophrenia. Schizophr Res 2:361–365
Kirch DG (1993) Infection and autoimmunity as etiologic factors in schizophrenia: a review an reappraisal. Schizophr Bull 19:355–370
Kirchner H, Kleinicke C, Digel W (1982) The whole-blood technique for testing production of human interferon by leukocytes. J Immunol Methods 48:213–219
Kita M, Shiozawa S, Yamaji M, Kitoh I, Kishida T (1991) Production of human a- and g-interferon is dependent on age and sex and is decreased in rheumatoid arthritis: a simple method for a large-scale assay. J Clin Lab Anal 5:238–241
Kolyaskina GI (1983) Blood lymphocytes in schizophrenia – immunological and virological aspects. Adv Biol Psychiatry 12:142–149

Maes M, Meltzer HY, Bosmans E (1994) Immune-inflammatory markers in schizophrenia: comparison to normal controls and effects of clozapine. Acta Psychiatr Scand 89:346–351

Maes M, Meltzer HY, Buckley P, Bosmans E (1995) Plasma-soluble interleukin-2 and transferrin receptor in schizophrenia and major depression. Arch Psychiatry Clin Neurosci 244:325–329

Maes M, Bosmans E, Ranjan R, Vandoolaeghe E, Meltzer HY, De Ley M, Berghmans R, Stans G, Desnyder R (1996) Lower plasma CC16, a natural anti-inflammatory protein, and increased plasma interleukin-1 receptor antagonist in schizophrenia: effects of antipsychotic drugs. Schizophr Res 21:39–50

Masserini C, Vita A, Basile R, Morselli R, Boato P, Peruzzi C, Pugnetti L, Ferrante P, Cazzullo CL (1990) Lymphocyte subsets in schizophrenic disorders. Schizophr Res 3:269–275

McAllister CG, Rapaport MH, Pickar D, Podruchny TA, Christison G, Alphs LD, Paul SM (1989) Increased numbers of CD5+ B lymphocytes in schizophrenic patients. Arch Gen Psychiatry 46:890–894

McAllister CG, van Kammen DP, Rehn TJ, Miller AL, Gurklis J, Kelley ME, Yao J, Peters JL (1995) Increases in CSF levels of interleukin-2 in schizophrenia: effects of recurrence of psychosis and medication status. Am J Psychiatry 152:1291–1297

McDaniel JS, Jewart RD, Eccard MB, Pollard WE, Caudle J, Stipetic M, Risby ED, Lewine R, Risch SC (1992) Natural killer cell activity in schizophrenia and schizoaffective disorder: a pilot study. Schizophr Res 8:125–128

Moises HW, Schindler L, Leroux M, Kirchner H (1985) Decreased production of interferon alpha and interferon gamma in leukocyte cultures of schizophrenic patients. Acta Psychiatr Scand 72:45–50

Monteleone P, Valente B, Maj M, Kemali D (1991) Reduced lymphocyte response to PHA and OKT3 in drug-free and neuroleptic-treated chronic schizophrenics. Biol Psychiatry 30:201–204

Müller N, Ackenheil M, Hofschuster E, Mempel W, Eckstein R (1991) Cellular immunity in schizophrenic patients before and during neuroleptic treatment. Psychiatry Res 37:147–160

Müller N, Hofschuster E, Ackenheil M, Eckstein R (1993) T-cells and psychopathology in schizophrenia: relationship to the outcome of neuroleptic therapy. Acta Psychiatr Scand 87:66–71

Nikkilä H, Müller K, Ahokas A, Miettinen K, Andersson LC, Rimón R (1995) Abnormal distributions of T-lymphocyte subsets in the cerebrospinal fluid of patients with acute schizophrenia. Schizophr Res 14:215–221

Nyland H, Naess A, Lunde H (1980) Lymphocyte subpopulations in peripheral blood from schizophrenic patients. Acta Psychiatr Scand 61:313–318

Pollmächer T, Hinze-Selch D, Mullington J, Holsboer F (1995) Clozapine-induced increase in plasma levels of soluble interleukin-2 receptors. Arch Gen Psychiatry 52:877

Pollmächer T, Hinze-Selch D, Mullington J (1996) Effects of clozapine on plasma cytokine and soluble cytokine receptor levels. J Clin Psychopharmacol 16:403–409

Preble OT, Torrey EF (1985) Serum interferon in patients with psychosis. Am J Psychiatry 142:1184–1186

Rapaport MH, Nelson DL, Paul SM (1989) Elevated levels of soluble interleukin 2 receptors in schizophrenia. Arch Gen Psychiatry 46:291–292

Rapaport MH, McAllister CG, Kirch DG, Pickar D (1991) The effects of typical and atypical neuroleptics on mitogen-induced T lymphocyte responsiveness. Biol Psychiatry 29:715–717

Rapaport MH, Fuller Torrey E, McAllister CG, Nelson DL, Pickar D, Paul SM (1993) Increased serum soluble interleukin-2 receptors in schizophrenic monocygotic twins. Eur Arch Psychiatry Clin Neurosci 243:7–10

Rapaport MH, McAllister CG, Kim YS, Han JH, Pickar D, Nelson DL, Kirch DG, Paul SM (1994a) Increased serum soluble interleukin-2 receptors in caucasian and korean schizophrenic patients. Biol Psychiatry 35:767–771

Rapaport MH, Lohr JB (1994b) Serum-soluble interleukin-2 receptors in neuroleptic-naive schizophrenic subjects and in medicated schizophrenid subjects with and without tardive dyskinesia. Acta Psychiatr Scand 90:311–315

Rogozhnikova OA, et al. (1993) Changes in B-cell immunity in the course of treatment of newly diagnosed schizophrenics. Zurnal Nevropatologii i Psichiatrii Imeni 93:65–68

Rothermundt M, Arolt V, Weitzsch C, Eckhoff D, Kirchner H (1996) Production of cytokines in acute schizophrenic psychosis. Biol Psychiatry 40:1294–1297

Rothermundt M, Arolt V, Peters M, Kirchner H (1998) Immunological dysfunction in schizophrenia. A systematic approach. Neuropsychobiology 37:186–193

Sasaki T, Nanko S, Fukuda R, Kawate T, Kunugi H, Kazamatsuri H (1994) Changes of immunological functions after acute exacerbation in schizophrenia. Biol Psychiatry 35:173–178

Schindler L, Leroux M, Beck J, Moises HW, Kirchner H (1986) Studies of cellular immunity, serum interferon titers, and natural killer cell activity in schizophrenic patients. Acta Psychiatr Scand 73:651–657

Smith RS, Maes M (1995) The macrophage-T-lymphocyte theory of schizophrenia: additional evidence. Med Hypotheses 45:135–141

Syvälahti EKG (1994) Biological factors in schizophrenia. Structural and functional aspects. Br J Psychiatry 164 [Suppl] 23:9–14

Tarazona R, Gonzalez-Garcia A, Zamzami N, Marchetti P, Frechin N, Gonzalo JA, Ruiz-Gayo M, van Rooijen N, Martinez-A C, Kroemer G (1995) Chlorpromazine amplifies macrophage-dependent IL-10 production in vivo. J Immunol 154:861–870

Villemain F, Chatenoud L, Galinowski A, Homo-Delarche F, Ginestet D, Loo H, Zarifian E, Bach J-F (1989) Aberrant T cell-mediated immunity in untreated schizophrenic patients: deficient interleukin-2 production. Am J Psychiatry 146:609–616

Wilke I, Arolt V, Rothermundt M, Weitzsch C, Hornberg M, Kirchner H (1996) Investigations of cytokine production in whole blood cultures of paranoid and residual schizophrenic patients. Eur Arch Psychiatry Clin Neurosci 246:273–278

Wright P, Gill M, Murray RM (1993) Schizophrenia: genetics and the maternal immune response to viral infection. Am J Med Genet 48:40–46

Correspondence: Dr. med. M. Rothermundt, Department of Psychiatry, University of Münster School of Medicine, Albert-Schweitzer-Strasse 11, D-48129 Münster, Federal Republic of Germany

On the myth of antibrain antibodies in schizophrenia – a critical overview

K. Schott, A. Batra, E. Richartz, R. Sarkar, M. Bartels, and **G. Buchkremer**

Universitätsklinik für Psychiatrie und Psychotherapie, Tübingen, Federal Republic of Germany

Summary

Antibrain antibodies are regarded as a hallmark of an autoimmune process in schizophrenia. The literature, however, is contradictory to a large extent. Furthermore, the older studies used now outdated methodology. The modern studies using Western and Immunoblot technique were greatly negative. Our own results on the occurrence of antibrain antibodies in schizophrenia were disappointing over the last years. In the present study we investigated antibody binding in serum from schizophrenic patients to the nuclear, membrane and water soluble protein fractions from human brain by Western and Immunoblot. There were no specific results. Therefore, we conclude that there is at present no proof of antibrain antibodies in schizophrenia.

Introduction

Antibrain antibodies are regarded as a hallmark of an autoimmune process in schizophrenia. Lehmann–Facius was the first in 1937 and 1939, who reported antibodies against "Hirnlipoid" in serum and CSF from schizophrenic patients. In 1985 DeLisi et al. reviewed the literature concluding that there is some evidence for the presence of antibrain antibodies in schizophrenia. However, the results were contradictory. Ehrnst et al. (1982) and Knight et al. (1990) reported negative findings. Furthermore, the methods applied in the older studies were now outdated tests like flocculation, agglutination, complement fixation and double diffusion in agar. Some authors used histological sections with indirect immunofluorescence staining. In the sixties Heath et al. (1967a) injected serum fractions from schizophrenics into the brain from monkeys and found EEG abnormalities. When they injected antibodies to brain tissue from sheep the monkeys exhibited catatonia and EEG abnormalities (Heath et al., 1967b). However, these results were not replicated and got into discredit during the last years.

An important point with these older studies is that the authors used complex antigen mixtures like histological sections and brain homogenates without making the attempt of characterising the target molecules. Today one should demand the proof of specific antigen – antibody systems in the CNS before claiming the occurrence of antibrain antibodies in schizophrenia. Even in 1994 Henneberg et al. reported an increased prevalence of antibrain antibodies in schizophrenia using immunofluorescence assay without specifying the target antigens.

Our own group has investigated the issue of antibrain antibodies in schizophrenia for years now using modern techniques as Western – and Immunoblot and ELISA. These techniques allow for an identification of target antigens when testing the binding of antibodies. In this study we report the search for serumantibodies against human brain proteins in paranoid schizophrenia.

Methods

Subjects

Controls were healthy blood donors without signs of mental disorder or autoimmune disease (4 men and 6 women, mean age 27.8 yrs, range 21–48 yrs). The patients with paranoid schizophrenia (5 men and 5 women, mean age 35.7 yrs, range 21–52 yrs) were diagnosed according to RDC (Spitzer et al., 1978). The patients were not controlled for medication.

Experimental procedures

Blood samples were obtained at routine venipuncture between 7:00 and 9:00 am, centrifuged and the serum frozen at −30°C until analysis.

Human brain was obtained at autopsy. The hippocampus was dissected rapidly and homogenized in ice – cold 0.25 M sucrose. Thereafter, the nuclear fraction was separated by centrifugation at 1000 g for 20 min. The pellet was washed in H_2O and the centrifugation was repeated. The sediment contains the nuclear fraction. The supernatant was centrifuged further at 57,000 g for 15 min. The sediment contains the membrane fraction. The supernatant contains water soluble proteins. The proteins were dissolved in 8% SDS and 2% DTT by heating at 80–90°C for 5 min and undissolved particles were spun down by centrifugation at 10,000 rpm. The SDS-proteins from the nuclear, membrane, and water soluble fractions were separated on 15% homogeneous microslab SDS – polyacrylamide gels as described by Poehling and Neuhoff (1980). The gels were stained for proteins with Coomassie/phosphoric acid according to

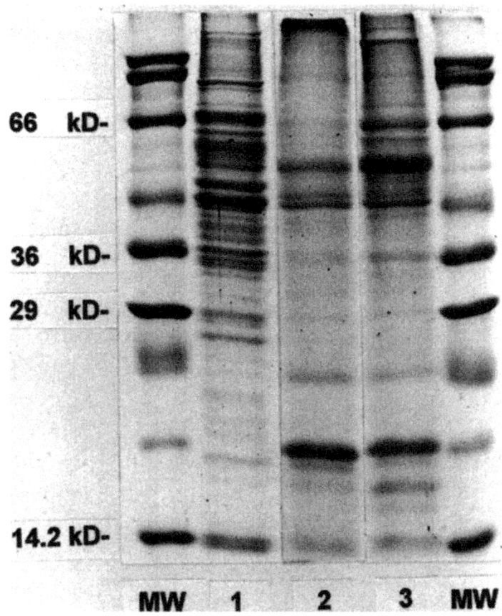

Fig. 1. Electropherogram of CNS proteins. From left: *MW* molecular weight markers; *1* water soluble proteins; *2* membrane proteins; *3* nuclear proteins

Neuhoff et al. (1985) or transferred onto NC sheets by semidry blotting at 300 mA for 3 hours.Immunostaining was developed according to Schott et al. (1988) using 4-chloro-1-naphthol as a substrate. Sera from controls and patients were diluted 1:50 and the second peroxidase conjugated antibodies were diluted at 1:1000. A cross-reaction of the second antibodies with the CNS antigens from human brain was ruled out before. The electropherograms of the CNS antigens are shown in Fig. 1.

Results

The results of the immunostaining on NC sheets are shown in Fig. 2 and 3.

In the case of antibody binding to nuclear proteins the IgM blot showed immunostaining of various proteins, which were identified as histone proteins by serological conditions. There was no difference of the staining pattern between controls and schizophrenic patients (Fig. 2). The IgG blot showed only a faint background staining but no specific bands.

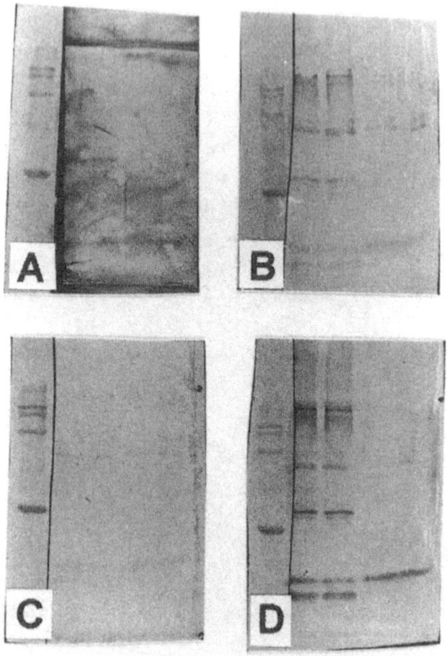

Fig. 2. Immunoblot of CNS proteins. **A** IgG blot from controls; **B** IgM blot from controls; **C** IgG blot from schizophrenic patients; **D** IgM blot from schizophrenic patients. Lanes from left: molecular weight markers, two lanes of nuclear proteins, and two lanes of membrane proteins

In the case of antibody binding to membrane proteins the IgM blot showed one immunostained protein band, but there was no difference between controls and schizophrenic patients (Fig. 2). This protein band could be identified later on as a histone protein. The IgG blot was empty.

In the case of antibody binding to water soluble proteins the IgM blot showed a pattern of high molecular weight immunostained proteins (Fig. 3). But there was no difference between controls and schizophrenic patients. The IgG blot was empty.

In summary, there were no specific immunostained CNS proteins in schizophrenia.

Discussion

Although there is a plenty of literature on antibrain antibodies in schizophrenia this issue is still under question. As outlined in the

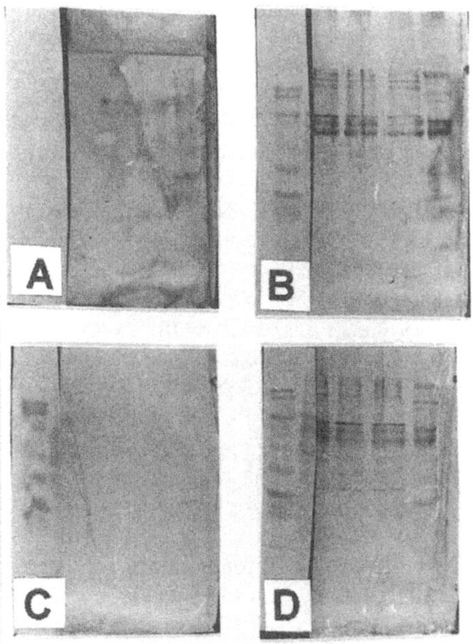

Fig. 3. Immunoblot of CNS proteins. **A** IgG blot from controls; **B** IgM blot from controls; **C** IgM blot from schizophrenic patients; **D** IgM blot from schizophrenic patients. Lanes from left: molecular weight markers, four lanes of water soluble proteins

Introduction the results were contradictory to a large extent. Our own group using Western Blot and ELISA was not able to detect autoantibodies to specific brain proteins from rat brain regions (Keller et al., 1995), neuropeptides (Batra et al., 1995), neurotransmitters (Schott and Batra, 1995), brain lipids (Schott et al., 1997), and synaptic membranes from human brain (Schott et al., 1998).

Only Plioplys et al. (1987) and Jankovic and Djodjijevic (1991) used comparable techniques like the Western Blot for characterising possible protein antigens from human brain. However, their results were not specific for schizophrenia either. Sundin and Thelander (1989) showed immunostaining of rat brain membrane proteins in schizophrenia but these results were not confirmed (Keller et al., 1995).

Our present study again shows clearly that there is no specific immunostaining of proteins from nuclear, membrane, and water soluble fractions of human brain in schizophrenia. As a result of our studies we conclude that there is at present no proof of antibrain antibodies in schizophrenia.

References

Batra A, Wegerer C, Bartels M, Günthner A, Mann K, Schott K (1995) Antibody reactivity in serum against neuropeptides in mental disorder. A dot – ELISA analysis. Neurol Psychiatry Brain Res 3:91–94

DeLisi LE, Weber RJ, Pert CB (1985) Are their antibodies against brain in sera from schizophrenic patients? Review and prospectus. Biol Psychiatry 20:110–115

Ehrnst A, Wiesel FA, Bjerkenstedt FB, Tribukait B, Jonsson J (1982) Failure to detect immunological stigmata in schizophrenia. Neuropsychobiol 8:169–171

Heath RG, Krupp IM, Byers LW, Lijekvist JI (1967a) Schizophrenia as an immunological disorder. II. Effects of serum protein fractions on brain function. Arch Gen Psychiatry 16:10–23

Heath RG, Krupp IM, Byers LW, Lijekvist JI (1967b) Schizophrenia as an immunological disorder. III. Effects of antimonkey and antihuman brain antibody on brain function. Arch Gen Psychiatry 16:24–33

Henneberg AE, Horter S, Ruffert S (1994) Increased prevalence of antibrain antibodies in the sera from schizophrenic patients. Schizophr Res 14:15–22

Jankovic BD, Djodjijevic D (1991) Differential appearance of autoantibodies to human brain S 100 protein, neuron specific enolase, and myelin basic protein in psychiatric patients. Int J Neur Sci 60:119–127

Keller A, Schott K, Batra A, Bartels M, Buchkremer G (1995) Brain antibodies in mental disorders. Neurol Psychiatry Brain Res 3:225–228

Knight JG, Knight A, Menkes DB, Muller PE (1990) Antibodies against brain septal region antigens specific to unmedicated schizophrenia? Biol Psychiatry 28:467–474

Lehmann–Facius H (1937) Über die Liquordiagnose der Schizophrenien. Klin Wochenschr 16:1646–1648

Lehmann–Facius H (1939) Serologisch – analytische Versuche mit Liquoren und Seren von schizophrenen bzw. atypischen Psychosen. Allg Z Psychiatr Grenzgeb 110:232–243

Neuhoff V, Stamm R, Eible H (1985) Clear background and highly sensitive protein staining with Coomassie blue dyes in polyacrylamide gels: a systematic analysis. Electrophoresis 6:427–448

Plioplys AV, Thibault S, Bouchard JP, Cockburn C, Hawkes R (1987) Anti-CNS antibodies in neurological and psychiatric disorders. J Neurol Neurosurg Psychiatr 50:1514–1521

Poehling HM, Neuhoff V (1980) One – and two – dimensional electrophoresis in microslab gels. Electrophoresis 1:90–102

Schott K, Batra A (1995) No antibody binding to serotonin in serum of patients with paranoid schizophrenia and major depressive disorder. Eur Psychiatry 10:372

Schott K, Jenner A, Wiethölter H, Schabet M, Stevens A (1988) Demyelination of sciatic nerve in acute experimental allergic neuritis of the rat. J Neuroimmunol 20:53–61

Schott K, Batra A, Richartz E, Sarkar R, Bartels M, Buchkremer G (1997) Serumantibodies to brain lipids in mental disorder. Neurol Psychiatry Brain Res 5:31–34

Schott K, Batra A, Richartz E, Sarkar R, Günthner A, Bartels M, Buchkremer G (1998) Antibrain antibodies in mental disorder: no evidence for antibodies against synaptic membranes. Neural Transm 105:517–524

Spitzer RL, Endicott J, Robins E (1978) Research diagnostic criteria. Arch Gen Psychiatry 35:773–782

Sundin U, Thelander S (1989) Antibody reactivity to brain membrane proteins in serum from psychiatric patients. Brain Behav Immun 3:345–358

Correspondence: PD Dr.med. K.–J. Schott, Universitätsklinik für Psychiatrie und Psychotherapie, Osianderstrasse 24, D-72076 Tübingen, Federal Republic of Germany

Schizophrenia: the teratogenic antibody hypothesis

P. Wright[1], **P. Laing**[2], **P.T. Donaldson**[3], and **R.M. Murray**[1]

[1] Department of Psychological Medicine, Institute of Psychiatry and King's College Hospital, De Crespigny Park, London,
[2] Peptide Therapeutics Group PLC, 321 Cambridge Science Park, Milton Road, Cambridge, and
[3] Institute of Liver Studies, King's College Hospital, Bessemer Road, London, United Kingdom

Summary

Disease discordance of 50% in monozygotic twins implicates genetic and environmental factors in the aetiology of schizophrenia. Autoimmune diseases are HLA associated and are thought to occur when a genetically predisposed individual is exposed to an essential, probably viral, environmental trigger. We have reported an excess of autoimmune diseases in the first degree relatives of schizophrenic patients, an excess of second trimester influenza infections in the mothers of schizophrenic patients, and a deficiency of HLA DRB1*04 alleles in the mothers of schizophrenic patients and in schizophrenic patients themselves. We have therefore hypothesised that maternal immunogenetic predisposition interacts with the influenza virus to cause neurodevelopmental lesions which manifest in adulthood as schizophrenia. Our work raises the possibility of an (immuno)genetic predisposition to schizophrenia at pedigree, maternal, and proband level.

Having a schizophrenic relative is the single most powerful risk factor for schizophrenia (Gottesman, 1991). However, disease discordance of about 50% in monozygotic twins (MZ) both confirms the contribution of genes to the aetiology of schizophrenia and implicates non-genetic environmental factors (Gottesman, 1991; Walker et al., 1991; Torrey, 1992). One hypothesis that may encompass these genetic and environmental aetiological contributions suggests that schizophrenia occurs because of genetic susceptibility to an unidentified environmental agent. An increased frequency of autoimmune diseases has been found in the relatives of schizophrenic patients (MacSweeney et al., 1978; Ganguli et al., 1987; DeLisi et al., 1991) and we have reported an excess of both insulin dependent diabetes mellitus (IDDM) and of thyrotoxicosis in the first degree relatives of 101 psychotic patients when compared to 115 controls (Gilvarry et al., 1996), and an excess of IDDM (but not of thyrotoxicosis) in the relatives of 121 DSM-III-R schizophrenic patients (Wright et al., 1996a) when compared to controls. Given the apparent contributions of both genes and environmental factors to schizophrenia, these reports are of particular interest because autoimmune diseases are thought to occur when a genetically predisposed individual is exposed to an essential, probably viral, environmental trigger. Furthermore, disease discordance of about 50% in MZ twins has been reported for several autoimmune diseases (Knight et al., 1992).

In this paper we discuss how immunogene-viral interactions may explain some of the epidemiological features of schizophrenia and will review the evidence implicating prenatal influenza virus as one environmental aetiological factor. We also consider potential mechanisms by which influenza might increase risk for schizophrenia and propose a model – the teratogenic antibody hypothesis – which may explain a proportion of those cases of schizophrenia that appear to be related to prenatal influenza exposure.

Schizophrenia – season of birth and prenatal influenza exposure

Schizophrenic patients are more likely to be born during late winter and spring than at any other time. Almost fifty studies report a 7%–15% excess of births of schizophrenic patients from approximately January until April in the northern hemisphere (Bradbury and Miller, 1985; Boyd et al., 1986; Pallast et al., 1994) while in the southern hemisphere Dalen (1975) and Parker and Neilson (1976) have reported that this season of birth effect was evident for schizophrenic females born from May to October. More recently, Welham et al. (1993) have demonstrated that in Queensland, Australia, immigrant schizophrenic patients born in Europe exhibit the northern hemisphere excess of births in the early months of the year, while Australian born schizophrenic patients have a similarly sized excess occurring later in the year during the southern hemisphere's winter and spring. Methodological artefacts – the so-called age incidence (Hare et al., 1974) and age prevalence (Lewis and Griffin, 1981; Lewis, 1989) effects – may partly explain these findings, but even when statistical corrections are applied (Lewis and Griffin, 1981; Pallast et al., 1994) or when age is calculated by month of birth (O'Callaghan et al., 1995), schizophrenic patients continue to exhibit a season of birth effect of the order of 10%. And it goes without saying that southern hemisphere studies such as that of Welham et al. (1993) are exempt from the methodological artefacts described above because the seasons of interest do not occur during the early part of the year. If not artefactual, what can explain the season of birth effect? The mothers of schizophrenic patients do not have unusual patterns of conception (Hare, 1976; McNeil et al., 1976; Buck and Simpson, 1978; Machon et al., 1983; Watson et al., 1984; Pulver et al., 1992) so we must look for some harmful environmental factor that adversely affects foetuses in utero during the winter months.

It was recently proposed that this teratogenic environmental factor may be exposure to influenza during the second trimester of foetal life. Almost a dozen ecological studies from Finland, Denmark, England and Wales, Japan, Scotland, and Australia have reported a positive association between prenatal influenza exposure and schizophrenia (Mednick et al., 1988; O'Callaghan et al., 1991; Sham et al., 1992; Fahy et al., 1992; Adams et al., 1993; Takei et al., 1994; Takei and Murray, 1994; McGrath et al., 1994; Kunugi et al., 1995). Some studies were equivocal (Adams et al., 1993; Kendall and Kemp, 1989) and others found no association (Torrey et al., 1991; Susser et al., 1994; Selten et al., 1994; Cannon et al., 1994), and of course ecological studies provide only epidemiological evidence of association between schizophrenia and maternal influenza infection. However, three case-control studies have been undertaken. Stober et al. (1992) reported that 9 of 55 (16.4%) mothers of schizophrenic patients recalled having an infection during the second trimester of gestation and four of these infections were influenza. Mednick et al. (1994) examined the obstetric records of the mothers of 25 schizophrenic offspring born in Helsinki in the wake of the 1957 influenza pandemic, and found that 86.7% of the women exposed to the pandemic during the second trimester of pregnancy had reported symptoms of influenza, compared with 20% who had been exposed during the other two trimesters. Finally, we have found that the mothers of 121 schizophrenic patients, 87 male and 34 female, recalled a significant excess of influenza during the second trimester compared with the combined first and third trimesters of their schizophrenic offspring's' gestations (11.6% vs. 1.7%; OR = 7.79, 95% CI = 1.63 to 50.81, df = 119, $x^2 = 9.64$, p = 0.002); we have further reported that schizophrenic patients whose mothers reported second trimester influenza had a significantly increased risk of obstetric complications (OCs) (OR = 4.84, Fisher's exact p = 0.011) and of a trend towards reduced birth weight (Wright et al., 1995). Parallels may be drawn between the suggestion that in utero exposure to influenza may cause schizophrenia – a central nervous system disorder – and (i) retrospective and prospective studies in Dublin which found that prenatal influenza exposure more than doubled the likelihood of neural tube defects (Coffey and Jessup, 1955, 1959) and (ii) Lynberg et al.'s (1994) report of an increased risk of neural tube defects in infants whose mothers recalled influenza with fever about the time of, and in the months following, conception. Thus it may be that in the case of schizophrenia maternal influenza causes impaired foetal neurodevelopment and that this in turn leads to OCs and reduced birth weight and, several decades later, to schizophrenia.

The teratogenic antibody hypothesis of schizophrenia

How could maternal influenza cause foetal brain disorder? It seems unlikely that influenza directly damages the foetal brain (Conrad and Schiebel, 1988) because influenza infection is not associated with a viraemia (Murphy and Webster, 1990; Laing et al., 1995). The foetus could suffer from alterations in maternal temperature, blood oxygen concentration, or acid/base or electrolyte balance brought about by influenza infection. However, such a mechanism would apply to most viral infections and schizophrenia appears to be associated only with prenatal exposure to influenza (O'Callaghan et al., 1994). Maternal antibodies can cause foetal brain disease and influenza A virus can both induce antibrain antibodies in experimental animals, and cause autoimmune neurological disorders in humans (see below). One possibility, therefore, is that some influenza associated agent of either maternal or viral origin is capable of transmitting the harmful effect of influenza virus from maternal respiratory tract epithelial cells to foetal neurons. What might this agent be?

Maternal IgG antibodies cross the placenta and neonatal diseases such as haemolytic disease of the newborn, thyrotoxicosis, myasthenia gravis, congenital heart block, pemphigus vulgaris, systemic lupus erythematosus and thrombocytopaenia are associated with the transplacental transfer of such maternal antibodies (Knight et al., 1992; Kaplan et al., 1992). Antibodies mediate these diseases directly – neonatal pemphigus vulgaris, for example, is caused by maternal antibodies acting against the 'pemphigus vulgaris antigen', a cell-adhesion molecule of the cadherin class (Amagai et al., 1991). Cadherins play a major role in neuronal cell-adhesion (Levitan et al., 1991) and it has been suggested that ectodermal ancestry may be a feature of all autoantigens (Tadmor et al., 1992), given that both skin and the central nervous system derive from embryonic ectoderm (Edelman, 1984). Minor physical anomalies (MPAs) involving structures of ectodermal origin are associated with schizophrenia (Waddington et al., 1990; Cannon et al., 1992) and it may be that MPAs and neurodevelopmental abnormalities reflect antibody-mediated perturbation of cell-adhesion in the developing foetal brain and integument respectively. Influenza may be associated with autoimmune neurological diseases including acute-disseminated encephalomyelitis following influenza-A infection (Boos and Esiri, 1986; Cohen and Lisak, 1987) and Guillain Barre' syndrome following vaccination against swine influenza (Safranek et al., 1991) in humans.

How could maternal influenza give rise to antibodies that cause foetal neurodevelopmental lesions? It may be that the influenza virus interacts with cell-surface sialoconjugates of CD56+ and neuronal cell adhesion molecules (NCAMs) on natural killer cells and T-lymphocytes, or with gangliosides on inflamed maternal respiratory epithelial cells, and leads to the production of autoantibodies cross reactive with these and similar brain molecules; there is extensive amino acid sequence homology at functionally important regions of the influenza HA1 protein and human embryonic cadherin for example (Blaschuk et al., 1990). We have therefore proposed that a small proportion of schizophrenia may represent an autoimmune disease (by proxy) in that antibodies elicited by the influenza virus in respiratory epithelial cells in immunogenetically predisposed women cross the placenta and immature foetal blood brain barrier during the second trimester of gestation and disrupt foetal neurodevelopment by cross reacting with brain antigens (Wright et al., 1993a,b).

Support for the teratogenic antibody hypothesis

We have attempted to provide both direct and circumstantial evidence in support of the teratogenic antibody hypothesis. Thus we have reported that autoantibodies against a 37 kDa brain protein may be detected in rabbits immunised with influenza-A virus (Laing et al., 1989) and we have preliminary data suggesting that the mothers of schizophrenic patients have higher titres of antibodies to an as yet uncharacterised 87–88 kDa brain protein than either their schizophrenic offspring or controls (Jameel, 1993). However, further attempts to replicate this finding have proved unsuccessful.

Autoimmune diseases are characterised by positive and negative associations with other autoimmune diseases and associations with human leucocyte antigens (HLA) – rheumatoid arthritis is associated with HLA DRB1*04 alleles, for example – and we have investigated such associations in schizophrenia in an effort to further support our theory. Almost 50% of schizophrenic patients have either autoimmune diseases or autoantibodies (Ganguli et al., 1987), a number of recent studies have reported increased rates of thyrotoxicosis and IDDM in the relatives of schizophrenic patients (DeLisi et al., 1991; Gilvarry et al., 1996; Wright et al., 1996a) and a negative association has been repeatedly demonstrated between schizophrenia and rheumatoid arthritis (Eaton et al., 1992). These findings suggest that a proportion of schizophrenia may be autoimmune in origin and thus may exhibit HLA association, and it has been postulated that the alleles associated with rheumatoid arthritis and with schizophrenia may be mutually exclusive (Knight et al., 1992). We therefore tested the hypothesis that there is genetic association of schizophrenia with HLA DRB1*04 alleles (Wright et al., 1993a) and found that DRB1*04 was present in 26.9% of schizophrenic patients and in 44.6% of comparison subjects (OR = 0.46, 95% CI = 0.25 to 0.81, df = 1, x^2 = 8.11, p = 0.004) (Wright et al., 1996b). Three other genotyping investigations of the DRB1 locus have found no associations (Zamani et al., 1994; Sasaki et al., 1994; Gibson et al., 1997) but Blackwood et al. (1996), using HLA serotyping rather than genotyping, reported that 29.7% of Scottish schizophrenic patients and 41.2% of comparison subjects had DR4. We have recently reported a trend for the preferential non-transmission of DRB1*04 alleles from heterozygous parents to schizophrenic offspring (16 of 23 alleles not transmitted, $x^2 = 3.5$, $p = 0.06$) (Wright et al., 1997) in 23 pedigrees and Schwab et al. (1998) has found similar results in a larger sample of 55 pedigrees using affected sib-pair analysis ($x^2 = 6.43$, $p = 0.012$).

It is believed that clearance of influenza infection depends on cytotoxic T lymphocyte killing of infected cells (Silver et al., 1992) and that HLA A1/A2/A3 are vital to the human immune response to influenza because they present the virus to T cells (DeBrino et al., 1993; Morrison et al., 1992). We therefore examined HLA A1, A2 and A3 in the mothers of 121 schizophrenic offspring, but found no overall differences in the frequency distributions of HLA A1/A2/A3 in these women compared with controls (Wright et al., 1995).

Conclusions

Our work raises the possibility of an (immuno)genetic predisposition to schizophrenia at pedigree, maternal, and proband level. We have reported an excess of autoimmune diseases in the first degree relatives of schizophrenic patients (Wright et al., 1996a) and a deficiency of HLA DRB1*04 alleles in the mothers of schizophrenic offspring and in schizophrenic patients who were unrelated either to these mothers or to each other (Wright et al., 1996b). The excess of second trimester influenza infections in the mothers of schizophrenic patients, and the association between these infections and obstetric complications and reduced birth weight (Wright et al., 1995), are in keeping with our hypothesis that maternal (and perhaps also proband) immunogenetic predisposition could interact with the influenza virus to cause neurodevelopmental lesions which manifest in adulthood as schizophrenia. Schizophrenia is associated with reduced fertility and one would have expected that the predisposing genes and the disease itself would have been progressively eliminated from the gene pool and the population. However, the relatives of schizophrenic patients are less susceptible to viral infections (Huxley et al., 1964; Carter and Watts, 1971) and the schizophrenia genotype may enhance resistance to infections such as influenza. Ironically, therefore, it may be that the genes responsible for the uniquely human devastation called schizophrenia may encode a biological advantage at all times other than when maternal influenza infection occurs within a discrete window of risk during mid-gestation.

References

Adams W, Kendell RE, Hare EH, et al. (1993) Epidemiological evidence that maternal influenza contributes to the aetiology of schizophrenia: an analysis of Scottish, English and Danish data. Br J Psychiatry 163:522–534

Amagai M, KlausKovtun V, Stanley JR (1991) Autoantibodies against a novel epithelial cadherin in pemphigus vulgaris, a disease of cell adhesion. Cell 67:869–870

Blackwood DH, Muir WJ, Stephenson A, Wentzel J, Ad'hiah A, Walker MJ, Papiha SS, StClair DM, Roberts DF (1996) Reduced expression of HLA B35 in schizophrenia. Psychiatr Genet 6:51–59

Blaschuk OW, Pouliot Y, Holland PC (1990) Identification of a conserved region common to cadherins and influenza strain A haemagglutinins. J Mol Biol 211:4, 679–682

Boos J, Esiri MM (1986) Viral encephalitis. Blackwell Scientific, Oxford, pp 1–365

Boyd JH, Pulver AE, Stewart W (1986) Season of birth: schizophrenia and bipolar disorder. Schizophr Bull 12:173–186

Bradbury TN, Miller GA (1985) Season of birth in schizophrenia: a review of evidence, methodology and etiology. Psychol Bull 98:569–594

Buck C, Simpson H (1978) Season of birth among the sibs of schizophrenics. Br J Psychiatry 132:358–360

Cannon M, Byrne M, Cotter D, et al. (1992) Further evidence for anomalies in the hand-prints of patients with schizophrenia: a study of secondary creases. Schizophr Res 13:179–184

Cannon M, Cotter D, Sham PC, et al. (1994) Schizophrenia in an Irish sample following prenatal exposure to the 1957 influenza epidemic: a case-controlled prospective follow up study. Schizophr Res 11:95

Carter M, Watts CA (1971) Possible biological advantages among schizophrenics' relatives. Br J Psychiatry 118:453–460

Coffey VP, Jessop WJE (1955) Congenital abnormalities – 6th series. Irish J Med Sci 349:30–46

Coffey VP, Jessop WJE (1959) Maternal influenza and congenital deformities. Lancet 2:935–938

Cohen JA, Lisak RP (1987) Acute disseminated encephalomyelitis. In: Aarli JA, Behan WMH, Behan PO (eds) Clinical neuroimmunology. Blackwell, Oxford, pp 192–213

Conrad AJ, Scheibel AB (1987) Schizophrenia and the hippocampus: the embryological hypothesis extended. Schizophr Bull 13:577–587

Dalen P (1975) Season of birth: a study of schizophrenia and other mental disorders. North Holland Publishing Company, Amsterdam

De Lisi LE, Boccio AM, Riordan H, et al. (1991) Familial thyroid disease and delayed language development in first admission patients with schizophrenia. Psychiatry Res 38:39–50

DiBrino M, Tsuchida T, Turner R, et al. (1993) HLA A1 and HLA A3 T cell epitopes derived from influenza virus proteins predicted from peptide binding motifs. J Immunol 11:5930–5935

Eaton WW, Hayward C, Ram R (1992) Schizophrenia and rheumatoid arthritis: a review. Schizophr Res 6:181–192

Edelman G (1984) Cell-adhesion molecules: a molecular basis for animal form. Sci Am 250:4, 80–91
Fahy TA, Jones PB, Sham PC (1992) Schizophrenia in Afro-Caribbeans in the UK following prenatal exposure to the 1957 A2 influenza epidemic. Schizophr Res 6:98–99
Ganguli R, Rabin BS, Kelly RH, et al. (1987) Clinical and laboratory evidence of autoimmunity in acute schizophrenia. Ann NY Acad Sci 496:676–685
Gibson S, Hawi Z, Straub R, Kendler K, Walsh D, Gill M (1997) Frequency of HLA-DR4 and HLA-DQB1*0602 in familial schizophrenics and controls. Ir J Med Sci 165:S2, 27
Gilvarry CM, Sham PC, Jones PB, Cannon M, Wright P, Lewis S, Bebbington P, Toone B, Murray RM (1996) Autoimmune diseases and psychosis: a case control family study. Schizophr Res 19(1):33–40
Gottesman II (1991) Schizophrenia genesis: the origins of madness. WH Freeman, New York, pp 104–132
Hare EH (1976) The season of birth of siblings of psychiatric patients. Br J Psychiatry 129:49–954
Hare EH, Price JS, Slater E (1974) Mental disorder and season of birth: a national survey compared with the general population. Br J Psychiatry 152:460–465
Huxley J, Mayr E, Osmond H, et al. (1964) Schizophrenia as a genetic morphism. Nature 204:4995, 220–221
Jameel SY (1993) Studies of virus induced autoimmunity to brain antigens in schizophrenia. Thesis, University of Nottingham
Kaplan C, Dehan M, Tchernia G (1992) Fetal and neonatal thrombocytopenia. Platelets 3:61–67
Kendell RE, Kemp IW (1989) Maternal influenza in the aetiology of schizophrenia. Arch Gen Psychiatry 46:878–882
Knight J, Knight A, Ungvari G (1992) Can autoimmune mechanisms account for the genetic predisposition to schizophrenia? Br J Psychiatry 160:533–540
Kunugi H, Nanko S, Takei N, et al. (1996) Schizophrenia following in utero exposure to 1957 influenza epidemics in Japan. Am J Psychiatry 152:450–452
Laing P, Knight JG, Hill JM, et al. (1989) Influenza viruses induce autoantibodies to a brain specific 37 kDa protein in rabbit. Proc Natl Acad Sci USA 86:1998–2002
Laing P, Knight J, Wright P, et al. (1995) Disruption of foetal brain development by maternal antibodies as an aetiological factor in schizophrenia. In: Mednick SA, Hollister JM (eds) Neural development in schizophrenia: theory and research. NATO Plenum Press, New York
Levitan IB, Kaczmarek LK (1991) The neuron: cell and molecular biology. Oxford University Press, New York
Lewis MS (1989) Age incidence and schizophrenia, part I. The season of birth controversy. Schizophr Bull 15:59–73
Lewis MS, Griffin TA (1981) An explanation for the season of birth effect in schizophrenia and certain other diseases. Psychol Bull 89:589–596
Lynberg MC, Khoury MJ, Lu X, et al. (1994) Maternal flu, fever, and the risk of neural tube defects: a population based case control study. Am J Epidemiol 140, 3:244–255

Machon RA, Mednick SA, Schulsinger F (1983) The interaction of seasonality, place of birth, genetic risk and subsequent schizophrenia in a high risk sample. Br J Psychiatry 143:383–388

MacSweeney D, Timms P, Johnson A (1978) Thyroendocrine pathology, obstetric morbidity and schizophrenia: survey of 100 families with a schizophrenic proband. Psychol Med 8:151–155

McGrath JJ, Pemberton M, Welham JL, et al. (1994) Schizophrenia and the influenza epidemics of 1954, 1957 and 1959: a southern hemisphere study. Schizophr Res 14:1–8

McNeil T, Kaij L, Dzierzykray-Rogalska M (1976) Season of birth among siblings of schizophrenics. Acta Psychiatr Scand 54:267–274

Mednick SA, Machon RA, Huttunen MO, et al. (1988) Adult schizophrenia following prenatal exposure to an influenza epidemic. Arch Gen Psychiatry 45:189–192

Mednick SA, Huttunen MO, Machon RA (1994) Prenatal influenza infection and adult schizophrenia. Schizophr Bull 20, 2:263–267

Morrison J, Elvin J, Latron F, et al. (1992) Identification of the nonamer peptide from influenza A matrix protein and the role of pockets of HLA A2 in its recognition by cytotoxic T lymphocytes. Eur J Immunol 22:903–907

Murphy BR, Webster RG (1990) Orthomyxoviruses. In: Fields BN, Knipe DM et al. (eds) Virology. Raven Press, New York

O'Callaghan E, Sham PC, Takei N, et al. (1991) Schizophrenia after prenatal exposure to 1957 A2 influenza epidemic. Lancet 337:1248–1250

O'Callaghan E, Sham PC, Takei N, et al. (1994) The relationship of schizophrenic births to 16 infectious diseases. Br J Psychiatry 165:353–356

O'Callaghan E, Cotter D, Colgan K, et al. (1995) Confinement of winter birth excess in schizophrenia to the urban born and its gender specificity. Br J Psychiatry 166:51–54

Pallast EGM, Jongbloet PH, Hubb M, et al. (1994) Excess seasonality of births among patients with schizophrenia and seasonal ovopathy. Schizophr Bull 20, 2:269–276

Parker G, Neilson M (1976) Mental disorder and season of birth: a southern hemisphere study. Br J Psychiatry 129:355–361

Pulver AE, Kung-Lee Yiang, Wolyniec PS, et al. (1992) Season of birth of siblings of schizophrenic patients. Br J Psychiatry 160:71–75

Safranek TJ, Lawrence DN, Kurland LT, et al. (1991) Reassessment of the association between Guillain-Barre syndrome and receipt of swine influenza vaccine in 1976–1977: results of a two state study. Am J Epidemiol 133:940–946

Sasaki T, Kuwata S, Dai XY, Nanko S, et al. (1994) HLA DR types in Japanese schizophrenica: analysis by group-specific PCR amplification. Schizophr Res 14:9–14

Schwab GS, Knapp M, Freymann J, Albus M, et al. (1998) The HLA DRB1 gene locus in schizophrenia: an association study in 55 families with linkage to chromosome 6p. Schizophr Res 29, 1–2:133

Selten JPCJ, Sleats JPJ (1994) Evidence against maternal influenza asa risk factor for schizophrenia. Br J Psychiatry 164:674–676

Sham PC, O'Callaghan E, Takei N, et al. (1992) Schizophrenia following prenatal exposure to influenza epidemics between 1939 and 1960. Br J Psychiatry 160:461–466

Silver ML, Guo H-C, Strominger JL, et al. (1992) Atomic structure of human MHC molecule presenting an influenza virus peptide. Nature 360:367–369

Stober G, Franzek E, Beckmann J (1992) The role of maternal infectious diseases during pregnancy in the aetiology of schizophrenia in offspring. Eur Psychiatry 7:147–152

Susser E, Lin SP, Brown AS, et al. (1994) No relation between risk of schizophrenia and prenatal exposure to influenza in Holland. Am J Psychiatry 151:922–924

Tadmor B, Putterman C, Naparstek Y (1992) Embryonal germ layer antigens: target for autoimmunity. Lancet 339:975–978

Takei N, Murray RM (1994) Prenatal influenza and schizophrenia. Br J Psychiatry 165:833–834

Takei N, Sham PC, O'Callaghan E, et al. (1994) Prenatal exposure to influenza and the development of schizophrenia: is the effect confined to females? Am J Psychiatry 151:117–119

Torrey FE (1992) Are we overestimating the genetic contribution to schizophrenia? Schizophr Bull 18:159–170

Torrey EF, Bowler AE, Rawlings R (1991) An influenza epidemic and the seasonality of schizophrenic births. In: Kurstat K (ed) Psychiatry and biological factors. Plenum Press, New York, pp 106–116

Waddington JL, O'Callaghan E, Larkin C (1990) Physical anomalies and neurodevelopmental abnormalities in schizophrenia: new clinical correlates. Schizophr Res 3:90

Walker E, Downey G, Caspi A (1991) Twin studies of psychopathology: why do the concordance rates vary? Schizophr Res 5:211–221

Watson CG, Kucala T, Tilleskjor C, et al. (1984) Schizophrenic birth seasonality in relation to the incidence of infectious diseases and temperature extremes. Arch Gen Psychiatry 41:85–95

Welham JL, Pemberton MR, McGrath JJ (1993) Schizophrenia: do seasonal birth rates vary between hemispheres? Schizophr Res 9:142–143

Wright P, Gill M, Murray RM (1993a) Schizophrenia: genetics and the maternal immune response to viral infection. Am J Med Genet 48:40–46

Wright P, Murray RM (1993b) Schizophrenia: prenatal influenza and autoimmunity. Ann Med 25:497–502

Wright P, Takei N, Rifkin L, Murray RM (1995a) Maternal influenza, obstetric complications and schizophrenia. Am J Psychiatry 152:1714–1720

Wright P, Donaldson P, Underhill J, Takei N, Rifkin L, Murray RM (1995b) Do maternal immunogenes mediate the 'schizophrenogenic' effect of prenatal influenza exposure? Psychiatr Genet 5(1):69

Wright P, Sham PC, Gilvarry CM, Jones P, Cannon M, Sharma T, Murray RM (1996a) Autoimmune disorders in the first degree relatives of schizophrenic and control subjects. Schizophr Res 20:261–267

Wright P, Donaldson PT, Underhill JA, Doherty DG, Choudhuri K, Murray RM (1996b) Genetic association of the HLA DRB1 gene locus on chromosome 6p21.3 with schizophrenia. Am J Psychiatry 153:1530–1533

Wright P, Dawson E, Donaldson PT, Underhill JA, Sham PC, Zhao J, Gill M, Nanko S, Owen MJ, McGuffin P, Murray RM (1998) A transmission/disequilibrium study of the DRB1*04 gene locus on chromosome 6p21.3 with schizophrenia. Schizophr Res 32:75–80

Zamani MG, De Hert M, Spaepen M, Hermans M, Marynen P, Cassiman JJ, Peuskens J (1994) Study of the possible association of HLA Class II, CD4 and CD3 polymorphisms with schizophrenia. Am J Med Genet 54:372–377

Correspondence: Dr. P. Wright, Department of Psychological Medicine, Institute of Psychiatry and King's College Hospital, De Crespigny Park, London SE5 8AF, United Kingdom

A preliminary analysis investigating immune activation in schizophrenia

M.H. Rapaport, M.P. Caliguiri, J.B. Lohr, D. Dolnak, S. Chavez, C. Kulesh, and **T. Hahn**

Department of Psychiatry, U.C. San Diego School of Medicine and the San Diego Veterans' Affairs Health System, San Diego, CA, USA

Summary

This paper is an interrum analysis of a longitudinal project designed to ascertain whether immune activation is a trait marker for a small subset of patients with the syndrome we call schizophrenia. We present data from over 70 schizophrenic patients and 50 matched normal volunteers. The patients were evaluated on 2 occasions at least 30 days apart. The preliminary analyses suggest that some patients with schizophrenia may have an immune modulated component to their disorder and this seems to be a relatively stable trait marker of the syndrome for these patients.

Introduction

This section of the paper establishes the rationale for the investigation of the immune system in schizophrenic patients by: (A) reviewing systemic illnesses which effect the central nervous system (CNS), (B) reviewing recent evidence implicating the immune system in schizophrenia, and (C) proposing a model linking TH1 cell activation to schizophrenia.

A. Systemic illness and central nervous system disturbance

There are considerable data suggesting that systemic illnesses may cause significant neuropsychiatric dysfunction. Tertiary syphilis is an historical example of systemic illness which once had profound neuropsychiatric effects. Although rare today, it was once common to see patients with dementia, poor impulse control, and paranoid behavior secondary to syphilis. Sydenham's chorea [1–4] and AIDS are two contemporary examples of infectious diseases which may have significant neuropsychiatric effects [1–4]. Sydenham's chorea is a post-streptococcal sequela associated with the production of cross-reactive antibodies which preferentially bind to brain tissue [1,4]. Patients with Sydenham's chorea may have abnormal movements, mood instability, and manifest symptoms of psychosis. Neuroleptics are the treatment of choice, and the illness recedes with the decrease in autoantibody titers [7]. Studies suggest that Sydenham's chorea may be associated with the development of obsessive-compulsive disorder in some individuals [4]. Autoimmune disorders such as multiple sclerosis and CNS lupus erythematosus have also been found to have significant neuropsychiatric consequences [1–4]. Central Nervous System lupus causes the development of psychotic symptomatology in 12 to 27% of patients and often presents in a manner that is indistinguishable from schizophrenia [4]. A number of patients with CNS lupus have been admitted to our psychiatric service with delusions and auditory hallucinations all of whom required neuroleptics until the immunosuppressive therapy ameliorated their symptoms.

B. Studies of immune dysfunction in schizophrenia

There were reports of leukocytosis in acute psychosis, of sera from schizophrenics precipitating brain extracts, and of blunted delayed hypersensitivity in schizophrenic subjects before neuroleptics were developed [5]. However, the results of the older studies of the immune system and schizophrenia were equivocal [4–8]. Limitations in the older literature are caused by: (1) differences in criteria used to diagnose schizophrenia (RDC vs. ICD vs. DSM) which alters the composition of the patients; (2) racial heterogeneity of study populations which affects HLA frequencies and other immune marker frequencies; (3) small sample sizes which limit the statistical power of many studies; (4) methodological differences in assays which may have yield different results for the same parameter; (5) failure to account for the effects of alcohol, smoking, and drugs of abuse on immune function; and (6) failure to account for the effects of neuroleptics on immune function.

Many newer studies investigating immune function and schizophrenia address these problems, and the data are more consistent. More recent investigations have found decreased IL–2 production in mitogen-stimulated lymphocyte cultures from schizophrenic subjects [9–16]. Several groups have also suggested that there may be an increase in serum IL–6 levels (a marker of macrophage activation) in some but not all schizophrenic patients [12,14,15]. Hornberg, Arolt, Wilke, Kruse and Kirchner reported that a subgroup of schizophrenic subjects had increased serum SIL–2R, decreased mitogen stimulated IL–2 and IFN–γ, and worse Strauss-Carpenter Scale scores [15]. The majority of work in this area has focused on studies of immune activation in patients who have acutely exacerbated [10–12,14–18]. These groups have described immune activation decreasing with treatment of psychosis [14–18]. Our group has been the main group investigating immune functioning in chronically ill schizophrenic patients and one consistent finding has been the elevation of serum SIL-2Rs. We and other have reported that SIL–2Rs is not markedly elevated in other psychiatric disorders [19,20] and are not subject to significant month-to-month variation [21], imply that further exploration of the peripheral cytokine network and schizophrenia is reasonable.

We believe that there are 2 distinct types of immune activation occur in patients with schizophrenia. One form of immune activation is general nonspecific immune activation which correlates with acute exacerbation's of psychosis and seems to resolve with stabilization on neuroleptics [13–18]. The other form of immune activation is due to chronic TH1 cell activation and is associated with intrinsic problems with movement disorders, and poor prognosis.

C. A model linking TH1 cell activation to schizophrenia

The SIL–2R is a 45 kd portion of a shed surface receptor (α–chain) which binds IL–2 with the same affinity as the low affinity surface IL–2 receptor [22]. SIL–2R are produced by activated T-cells, B-cells, and activated macrophages. SIL-2Rs are markedly elevated during acute infections, leukemias, transplant rejections, and autoimmune disorders and may be mildly increased by smoking [22]. IL–2 stimulates the production of SIL-2Rs which are thought to be a component of the negative feedback loop to modulate IL–2 [22].

Cytokines are proteins which when first discovered were found to be important in the regulation of the immune system, but recently cytokines have been implicated in the modulation of a variety of processes including fever, sleep, CRF and neurotransmitters [23–31]. Studies show in vivo synthesis of cytokines occurs within the brain [24,26,28], that there is potential communication via circumventricular organs (which have no blood-brain barrier) between the periphery and the CNS , and that there are active transport systems for cytokines from the periphery into the CNS [30,31]. Activated T-cells have been demonstrated to cross normal blood brain barriers in autoimmune diseases such as multiple sclerosis, and there are now reports of similar findings in some, but not all, schizophrenic patients [32]. One study in child psychiatry evaluated CSF cytokine levels in schizophrenia, OCD, and ADHD. The majority of schizophrenic patients had detectable levels of IL-2 and IL-4. However, 3/22 schizophrenic patients had detectable IFN-γ levels and two of these patients had very high IFN-γ levels [33]. These findings may identify a minority of schizophrenic patients with an immune modulated disorder and suggests that there are variety of mechanisms by which peripheral immune activation could modulate CNS function.

Developments in immunology show that patterns of cytokine secretion are linked to specific T-helper lymphocyte repertoires: type 1 CD4+ helper T-cells (TH1) elaborate IFN–γ and IL–2 but not IL–4 or IL–5, while type-2 helper-T cells (TH2) preferentially secrete IL–4 and IL–5 but not IFN–γ or IL–2 [34–36]. This is a positive mediator for TH1 cells which are associated with delayed type hypersensitivity and other effector functions related to macrophage interaction and T-cell mediated immunity; IL–10 seems to stimulate TH2 cells which modulate atopic and allergic reactions and helminthic infections [34–36]. This differentiation of the functions of cytokines in both immune and CNS modulation helps to increase our understanding of immune findings in schizophrenia, and it puts the findings in an understandable context.

Our work and the work of other researchers, found that serum SIL–2R is markedly elevated (greater than 2 SD beyond the mean of controls) in 30 to 40% of schizophrenic patients but not in other psychiatric disorders. Some schizophrenic subjects also have been found to have diminished IL–2 production in response to mitogen exposure [10,11], increased peripheral IL–6 levels [12,14,15], increased CSF IL–2 levels [37], and increased T-cells in their CSF despite intact blood brain barriers [32]. This pattern of immune activation is consistent with TH1 cell activation. Increased IL–2 levels may be important because IL–2 has been demonstrated to modulate central dopaminergic and opioid tone [26,27]. In animal models it has been shown that IL–2 can be synthesized in the striatum and stimulates dopamine activity in the caudate, putamen, and nigro-striatal areas [26]. Dr. Kirk Denicoff and colleagues have found that some cancer patients undergoing LAK therapy have very high CSF IL–2 levels, and they develop severe neuropsychiatric dysfunction, including psychotic symptoms, which are remedied by Haloperidol treatment [38].

The importance of motor dysfunction in schizophrenia has been recognized since Kraepelin; however, the study of spontaneous movement disorders of putative extrapyramidal origin (Parkinsonism and dyskinesia) and volitional movement disorders (catatonia, mannerisms, and stereotypies) have become increasingly prominent since the introduction of neuroleptics Recent studies suggest that spontaneous dyskinesia may have an illness-based etiology particularly in schizophrenic patients with intellectual impairment and negative symptoms. However, we demonstrated in previous studies [39] that there was an association between TD and elevated serum SIL–2R levels, and a relationship between motor abnormalities and serum SIL–2R levels. We studied motor instability in 32 schizophrenic subjects to determine whether motor abnormalities were associated with increased SIL-2Rs (2 SD beyond the mean of controls). The force instability score was greater for the 13 patients with high SIL-2Rs levels (mean = 3.7 + 1.4) vs. the 19 patients with low SIL–2R levels (2.9 + 0.6: U = 7.3, Z = 2.65, p = 0.008). In a second analysis, we examined the relationship between degree of force instability and SIL–2R levels for the cohort of 32 patients. We found a correlation: (r = .54, df 31, p = .001), and this correlation held even when we looked only at the small cohort of neuroleptic naive patients in the sample, (r = 0.73, df 9, p = 0.02) [40]. These findings suggest that disturbance in the control of muscle force is not only pathognomonic to neuroleptic-induced hyperkinesia but may represent a marker of vulnerability

Based on a synthesis of the data presented, we hypothesized that TH1 cell activation causes increased IL–2 production; and that increased IL–2 production, possibly in concert with decreased dopaminergic feedback inhibition from the mesocortical dopamine system, increases mesolimbic and nigrostriatal dopamine system activity to cause the positive symptoms of psychosis and motor dysfunction. We postulate that a subgroup of schizophrenic patients have TH1 cell activation driving their illness. These patients will have more psychopathology, more objective signs of movement disorders, and will have a worse clinical course. If this model or a variant of it is correct, we will be able to subtype biologically some patients with schizophrenia.

Methods

Criteria for selection of schizophrenic subjects

Participants had a SCID interview, PANS, SANS, SAPS, AIMS. Subjects must receive a DSM-IV diagnosis of schizophrenia. Subjects may meet lifetime criteria for other Axis I disorders, but these disorders must be in remission. The diagnosis of TD was using the Schooler and Kane criteria. We also performed the Scale of Functioning (SOF).

Evaluation before entrance into the study

All subjects had complete medical histories, physical examinations, and screening laboratory work. Age of onset, clinical course, history of medication exposure, family history of psychiatric illness or autoimmune illness, state of illness, and frequency of alcohol and tobacco have been gathered as part of the evaluation. Subjects were excluded if they had current infections, currently active allergies, cancer or history of cancer, take medications that affect the immune system, or met diagnostic criteria for alcohol or substance dependence.

Preparation and storage of samples

Between 7:30 and 9:30 A.M., blood were drawn from non-fasting subjects both serum and samples for lymphocyte culture will be drawn in EDTA containing tubes. Serum were obtained from clotted blood by centrifugation at 500 g for 10 minutes. Samples were be frozen at –80°C.

Soluble interleukin 2 receptor (SIL-2R) assay

The SIL-2Rs will be measured using the Cellfree SIL-2R kit manufactured by T-Cell Sciences, Cambridge, Massachusetts. This is a commercial sandwich ELISA which uses 2 monoclonal antibodies which react with different epitomes of the SIL-2R. Intra-and inter-assay CV are 5 and 10 percent.

IL-2, IL-4, IFN-γ, IL-10 Assays

Assessment of cytokines will be done by commercially available, competitive enzyme immunoassays purchased from CytoImmune Sciences of College Park, Maryland. The procedure relies on the competition between conjugated cytokine and specific cytokines in standards or samples for a limited number of specific antibody binding sites. Data are analyzed by computer-assisted logit-log transformation of the resultant ODs using an automated ELISA reader. The limit of sensitivity of the assay is 0.1 ng/ml, with intra- and inter-assay CVs of 7% and 15%, respectively. An aliquot of frozen serum from a healthy, normal control who has donated a large serum sample will be used as a batch control in each assay.

Quantitative assessment of bradykinesia

A platform instrumented with a rotation sensor attached to a handle (electrogoniometer) is used to record wrist rotation. Low voltage current from a bridge amplifier is passed through the electrogoniometer. Changes in wrist rotation results in a proportion change in the resistance to that current and effected a proportional change in voltage. The voltage output from the electrogoniometer is digitized at 100 samples/second using a 12-bit analog-digital board (Real Time Devices, PA) installed into a microcomputer. Custom software will be used to convert the digitized signal to a large cursor displayed on the monitor. The position of the cursor along the horizontal plane is calibrated in degrees of rotation. In addition to the rotation-based cursor, target boxes are displayed on the monitor, also along the horizontal plane and are located at 25 or 45 degrees left of midline for right wrist flexion, and right of midline for left wrist flexion. Subjects sit facing a computer monitor while they hold the electrogoniometer handle with one hand. When a target box appeared on the screen, subjects flexed their wrist "as quickly and as accurately as possible" to reach the target. The target remains displayed for 2 seconds with an inter-stimulus interval of 2 seconds. If the subject needs clarification, instructions are repeated with emphasis on "speed of movement". Thirty-two trials or movements, consisting of 16 trials for each of two randomly presented target locations are administered for each hand, for a total of 64 movements.

Because the movement task involves rapid ballistic wrist flexion the task may be considered an example of open loop control, that is, movements that do not depend or make use of proprioceptive feedback. However, since the task also involves movement to a visually displayed target, visual feedback processes may potentially interact with the open loop processes in the production of wrist flexion.

Current theories of motor control suggest that an important role of feedback is to provide information about the state of the motor system *prior* to movement, such that the motor system programs the endpoint of the movement based on this information [41–43].

Figure 1 shows waveforms for displacement (waveforms A and C) and velocity (waveforms B and D) from two subjects during execution of the motor task. The waveforms selected to demonstrate normal (B and D) and abnormal (A and C) adjustments to velocity during rapid movements to targets varying in distance. The figure shows that abnormal performance is characterized by velocity peaks for the more distant target (45 degrees) that are indistinguishable from the velocity peaks for the nearer target (25 degrees). In contrast, normal performance shows increased velocity for the 45 degree target relative to the 25 degree target. Although the pattern of increasing velocity to meet the demands of the task is demonstrated in the figure, the "law" does not hold for every trial.

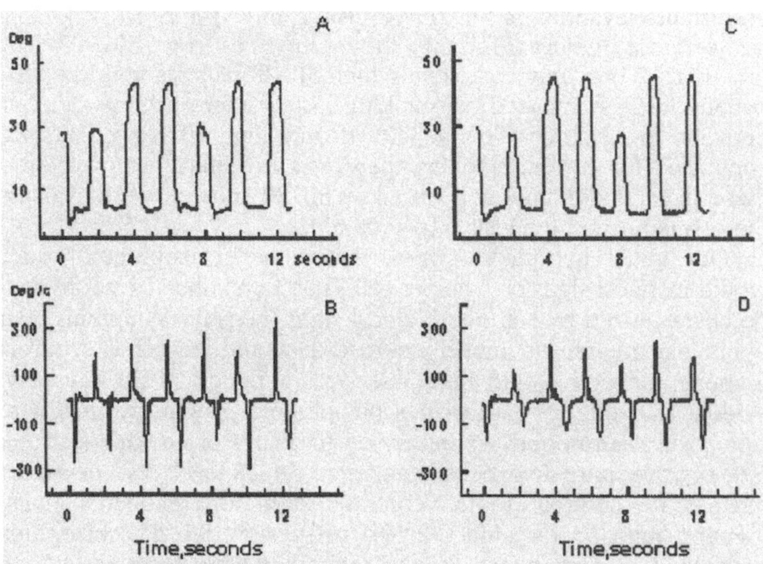

Fig. 1. Exemplary rotation and velocity waveforms showing systematic changes in velocity for the normal condition (right) and failure to adjust velocity in the abnormal condition (left)

Also noteworthy in this figure are the negative velocity peaks indicating rapid movements back to the starting point. Interestingly, at least in the present example, negative velocities appear to be scaled to distance in the abnormal condition. This is not inconsistent with our contention that neuromotor deficits are associated with failures to scale velocity to meet the demands of the task since the task did not require the subject to return to the starting point with any degree of accuracy. Thus, tasks which include a specific target or end-point presume that velocity is programmed prior to movement, whereas in the absence of a target, movement velocity is not a control parameter [41].

Data from 32 trials from each hand are incorporated into separate linear regression analyses to obtain the slope coefficient of the peak velocity/peak rotation linear function. The slope coefficient serves as the index of velocity scaling and is derived from the equation: $y = mx + b$, where y represents peak instantaneous velocity in degrees/second; x, rotation in degrees; m, the slope coefficient; and b, the intercept constant.

Results

We have assayed samples from 76 schizophrenic patients at two time points and over 50 controls. We demonstrate consistently elevated levels for the SIL-2Rs in some patients. We detected serum levels of IL-2 and interferon γ for the majority of patients. However, we did not detect consistent levels of IL-6, IL-4 or IL-10; We performed analyses to determine whether age, gender, or smoking history accounted for consistent elevations of SIL-2Rs at Visit 1 and Visit 2. 3×2 χ^2 analyses were not significant for any of these analyses ($p = 0.599$; $p = 0.422$, $p = .154$). When only consistently high SIL-2R patients and low SIL-2R patients were contrasted on smoking history there was no difference between the 2 groups ($p = 0.340$). If only high SIL-2R patients were contrasted for current smoking there was no effect but over 80% of these patients did smoke. If only low SIL-2R subjects were evaluated, there was a trend for Visit 1 ($p = 0.067$).

Our initial hypotheses were: that some schizophrenic patients would manifest signs of T-helper cell Type I immune activation; would be characterized by soft neurological signs, negative symptoms, extra-pyramidal side effects, movement disorders, and the deficit syndrome; immune function would be stable over a period of 30 days. As is evident in Table 2, Caucasian schizophrenic patients were different from Caucasian normal volunteers on measures of immune activation. 50% of schizophrenic patients manifested SIL-2R levels >SD beyond the mean of the controls. Table 1 contrasts serum and mitogen stimulated immune function for high (>2 SD) versus low SIL-2R schizophrenic patients. Anti-smooth muscle autoantibodies were the only autoantibody consistently elevated with 40% of patients being autoantibody positive. Autoantibody status did not correlate with "high" versus "low" immune status. Analysis of the relationship between clinical

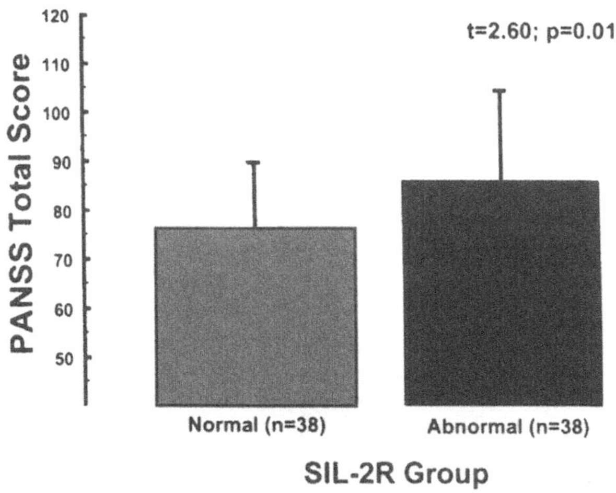

Fig. 2

Table 1. Visit 1 mean cytokine and cytokine-receptor levels for "high" and "low" schizophrenic patients and normal controls

	High (SD)	Low (SD)	Normal controls
SUP IL-2	23.9 (41.0)	23.8 (48.6)	11.91 (13.72)
SUP IL-10	401.1 (474)	***583.7 (344.9)***	773.87 (744.55)
SUP IFN-γ	340.9 (708.6)	***466.8 (945.0)***	151.39 (412.74)
SER SIL-2R	***1,490.6 (750.1)***	486.5 (151.6)	511.25 (127.58)
SER IL-2	***48.9 (47.0)***	30.4 (7.2)	29.70 (7.01)
SER IL-4	7.2 (5.4)	8.9 (7.5)	Not dectable
SER IL-6	***110.8 (307.2)***	36.7 (62.2)	37.08 (99.56)
SER SIL-6R	2,048 (5,234.5)	2,637 (188.4)	188.14 (82.36)
SER IL-10	12.9 (11.7)	20.4 (8.7)	Not detectable
SER IFN-γ	21.0 (3.8)	19.98 (2.5)	24.96 (26.79)

functioning and immune activation is incomplete. The mean clinical ratings for "high" vs. "low" immune activation were: BPRS-18 48.8 (10.3) vs. 45.4 (9.5), SAPS-Tot 8.5 (4.4) vs. 7.1 (3.4), SANS-Tot 10.5 (5.5) vs. 9.4 (5.5); PANS-Tot 85.7 (19.4) vs. 76.6 (16.1) ($p = 0.01$); SOF-Tot 35.2 (6.4) vs. 38.0 (7.8) ($p = 0.22$). We did not find a correlation between increased immune activation and the Deficit Syndrome. Other analyses are in progress.

Although there were dramatic differences between cytokine levels only SIL-2R levels were statistically significant at $p < 0.05$.

Table 2. Visit 2 mean cytokine and cytokine-receptor levels for "high" and "low" schizophrenic patients

	High	Low
SUP IL-2	***17.2 (9.3)***	***30.1 (28.9)****
SUP IL-10	***369.8 (322.7)***	***960.3 (1,691.7)*****
SUP IFN-γ	***136.2 (289.9)***	***337.4 (625.6)******
SER SIL-2R	***877.7 (483.3)***	***422.1 (109.8)*******
SER IL-2	111.4 (40.1)	82.4 (41.1)
SER IL-4	3.5 (.99)	2.4 (.54)
SER IL-6	63.2 (61.9)	43.8 (32.2)
SER IL-10	9.9 (4.6)	8.2 (4.5)
SER IFN-γ	11.6 (27.9)	5.5 (6.2)

*F = 3.73 P = .059 **F = 2.74 P = .103 ***F = 3.59 *P* = .063 ****F = 4.31 P = .045

Sixty-five percent (24/38) of individuals with SIL-2Rs >2 SD beyond the mean of controls maintained the status at Visit 2. As is evident in Table 2, schizophrenic patients with increased SIL-2Rs at Visit 2 were distinctly different from those schizophrenic patients with normal SIL-2R levels.

Because of lack of funds we did not perform SIL-6R assays at Visit 2, and only performed normal control SIL-2R assays at Visit 2. A further constraint was that detectable levels of IFN-γ, IL-4, IL-6, and IL-10 varied significantly between Visit 1 and Visit 2. Changes in detectable levels did not correlate with clinical variables. *The McNamara Test demonstrates that a significant number of immune activated patients have a consistent immune presentation at both Visit 1 and Visit 2.* When we compare and contrast the clinical characteristics of patients "high" SIL-2Rs vs. "low" SIL-2Rs: BPRS-18 46.6 (10.1) vs. 43.5 (9.6); SAPS – Tot 8.8 (4.4) vs. 7.8 (3.8); SANS – Tot 10.6 (5.1) vs. 10.1 (5.7); PANS – Tot 82.1 (17.3) vs. 80.1 (17.9); SOF – Tot 35.7 (6.7) vs. 38.8 (6.7) ($p = 0.19$).

We have previously published data suggesting a relationship between the presence of and an objective measure of the risk of developing tardive dyskinesia electroergomonically measured tremor. As illustrated in Fig. 3, we have determined that bradykinesia is correlated with serum IL-2 levels.

Discusson

As is the case with all science, actual data do not precisely match hypothesized expectations. In general these data suggest that some patients with schizophrenia manifest immune activation which may be

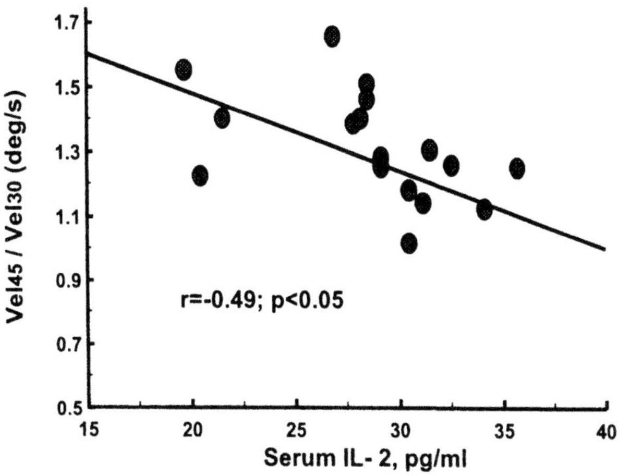

Fig. 3

best conceptualized as possibly related to TH1 activation at this time. (It is clear as our knowledge of immunology and autoimmunity evolve, that the triggers for TH1-like and TH2-like cytokine dysregulation associated with autoimmune diseases maybe the overproduction or underproduction of modulatory factors/cytokines such as TNF-α or IL-2.) We consistently find that some schizophrenic patients do possess serum SIL-2Rs that are 2 SD > the mean of controls and that this cannot be explained by medications, smoking, or other clinical factors. We find a pattern of purified lymphocyte mitogen-stimulated cytokine production which is suggestive of TH1-like activation. Yet, it was only statistically significant at the second evaluation point. We postulate that 2 factors contribute to this: first, we and others feel that whole blood assay results are possibly more representative of the internal biological milieu and give more robust levels of cytokine production; and second, at the second sampling point only 65% of the initial cohort of high patients remained high and these individuals seem to be more clearly differentiated from the low SIL-2R patients. At this time we do not know why some patients shifted groups since none of the clinical variables evaluated accounted for this transition from high to low. Only one subject switched from low to high and this person developed manifestations of an acute infection after the sample was drawn. The data presented at Table 2 includes all subjects evaluated at time 2, however, when we evaluated patients who were low at both time points and the 14 patients who went from the high group to the low group, we did not see any statistically or clinically significant differences in cytokine profiles. The clinical variables most closely correlated with immune activation continue to be motor dysfunction. However, as seen in Fig. 2, patients in the high group tended to have more psychopathology and were less functional, although SOF differences did not reach statistical significance at this time. Part of this may lack of significance be a floor effect since we recruited a large number of clinically ill conservatorized patients into this study. These findings suggest to us that further careful evaluation of this cohort is necessary to ascertain: (1) the etiology of immune activation in these patients, (2) whether these signs of peripheral immune activation are indicative of CNS immune activation (3) the stability of TH1-like activation longitudinally, (4) the relationship between immune activation, measures of psychopathology and motor dysfunction

Acknowledgements

This work is supported by a VA Merit Review Board Grant, MH 30914-21, MH 49746-02 and the VISN 22 VA MIRECC

References

1. Swedo S, Rapoport J, Cheslow B, et al. (1989) High prevalence of obsessive-compulsive symptoms in patients with Sydenham's Chorea. Am J Psychiatry 146:246–249
2. Sindic CJ, Laterre EC (1991) Oligoclonal free kappa and lambda bands in the cerebrospinal fluid of patients with multiple sclerosis and other neurological diseases. An immunoaffinity-mediated capillary blot study. J Neuroimmunol 33:63–72
3. Van Dam AP (1991) Diagnosis and pathogenesis of CNS lupus. Rheumatol Int 11:1–11
4. Rapaport MH, McAllister CG (1991) Neuroimmunologic factors in schizophrenia. In: Gorman J, Kertzner R (eds) Psychoimmunology update. American Psychiatric Press, Washington, DC, pp 31–47
5. Lehman-Facius H (1939) Serologisch-analytisch Verche mit Liquoren und Seren von Schizophrenien. Alg Z Psychiatr 110:232–243
6. Johnstone E, Whaley K (1975) Antinuclear antibodies in psychiatric illness: their relationship to diagnosis and drug treatment. Br Med J 724–725
7. Kronfol Z, Turner R, Nasrallah H, et al. (1984) Leukocyte regulation in depression and schizophrenia. Psychiatry Res 13:13–18
8. Liedman R, Prilipko L (1978) The behavior for lymphocytes in schizophrenia. Original Article Series XIV, No 5:365–377
9. Villemain F, Chatenoud L, Guillibert E, et al. (1987) Decreased production of interleukin-2 in schizophrenia. Ann NY Acad Sci 496:669–675
10. Villemain F, Chatenoud L, Galinowski A, et al. (1989) Aberrant T-cell-mediated immunity in untreated schizophrenic patients: deficient interleukin-2 production. Am J Psychiatry 146:609–616
11. Ganguli R, Rabin BS, Belle SH (1989) Decreased interleukin-2 production in schizophrenic patients. Biol Psychiatry 26:427–430

12. Shintani F, Kanba S, Maruo N, et al. (1991) Serum interleukin-6 in schizophrenic patients. Life Sci 49:661–664
13. Maes M, Meltzer H, Bosmans E (1994) Immune-inflammatory markers in schizophrenia: comparison to normal controls and effects of clozapine. Acta Psychiatr Scand 3:4–6
14. Katila H, Appelberg B, Hurme M, et al. (1995) Plasma levels of interleukin-1 b and interleukin-6 in schizophrenia, other psychoses, and affective disorders. Schizophr Res 12:29–34
15. Hornberg M, Arolt V, Wilke I, et al. (1995) Production of interferons and lymphokines in leukocyte cultures of patients with schizophrenia. Schizophr Res 15:237–242
16. Maes M, Meltzer HY, Buckley P, Bosmans E. Plasma-soluble interleukin-2 and transferrin receptor in schizophrenia and major depression. Eur Arch Psychiatry Clin Neurosci 244:325–329.
17. Muller N, Manfred A, Hofschuster E, Mempel W, Eckstein R (1991) Cellular immunity in schizophrenic patients before and during neuroleptic treatment. Psychiatry Res 37:147–160
18. Maes M, Bosmans E, Calabrese J, Smith R, Meltzer HY (1995) Interleukin-2 and interleukin-6 and mania: effects of neuroleptics and mood stabilizers. J Psychiat Res 29(2):141–152
19. Rapaport MH (1994) Immune parameters in euthymic bipolar patients and normal volunteers. J Affect Disord 32:149–156
20. Rapaport MH, Stein MB (1994) Serum interleukin-2 and soluble interleukin-2 receptor in generalized social phobia. Anxiety 1:50–53
21. Maes M, Stevens W, Scharpe S, et al. (1994) Seasonal variation in peripheral blood leukocyte subsets and in serum interleukin-6, and soluble interleukin-2 and -6 receptor concentrations in normal volunteers
22. Rubin LA, Nelson DL (1990) The soluble interleukin-2 receptor: biology, function and clinical application. Ann Intern Med 113:619–627
23. Barbanel G, Ixart G, Szafarczyk A, et al. (1990) Intrahypothalamic infusion of interleukin-b increases the release of corticotropin-releasing hormone (CRH 41) and adrenocorticotropic hormone (ACTH) in free-moving rats bearing a push-pull cannula in the median eminence. Brain Res 516:31–36
24. Nistico G, De Sarro G (1991) Is interleukin-2 a neuromodulator in the brain? TINS 14:146–150
25. Besedowsky H, Del Rey A, Sorkin E et al. (1983) The immune response evokes changes in brain noradrenergic neurons. Science 221:564–566
26. Lapchak PA (1992) A role for interleukin-2 in the regulation of striatal dopaminergic function. Neuroreport 3:165–168
27. Lapchak PA, Araujo DM (1993) Interleukin-2 regulates monoamine and opioid peptide release from the hypothalamus. Neuroreport 4:303–306
28. Lapchak PA, Araujo DM, Quirion R, et al. (1991) Immunoradiographic localization of interleukin 2-like immunoreactivity and interleukin 2 receptors (Tac antigen-like immunoreactivity) in the rat brain. Neuroscience 44:173–184
29. Saper CB, Breder CD The neurologic basis of fever. N Engl J Med 330:1880–1886
30. Banks WA, Ortiz L, Plotkin R et al. (1991) Human interleukin (IL) 1a, murine IL-1a and murine IL-1b are transported form blood to brain in the mouse by a shared saturable mechanism. J Pharmacol Exp Ther 259:988–996

31. Waguespack P, Banks W, Kastin A (1993) Interleukin-2 does not cross the blood-brain barrier by a saturable transport system. Brain Res Bull 34:103–109
32. McAllister C, Konicki E, Kirch D, et al. (1991) Increased percentage of T-lymphocytes in CSF of schizophrenic patients. Schizophr Res 4:373
33. Mittleman BB, Castellanos FX, Jacobsen LK, Rapoport JL, Swedo SE, Shearer GM (1997) Cerebrospinal fluid cytokines in pediatric neuropsychiatric disease. J Immunol 159:2994–2999
34. Durham S, Ying S, Varney V, et al. (1992) Cytokine messenger RNA expression for IL-3, IL-4, IL-5, and granulocyte/macrophage-colony stimulating factor in the nasal mucosa after local allergen provocation: relationship to tissue rosinophilia. J Immunol 148:2390–2394
35. Psicopoulos A, Hamid W, Varney V (1992) Preferential messenger RNA expression of TH1 type cells (IFN-gamma, IL-2+) in classical delayed-type (tuberculin) hypersensitivity reactions in human skin. J Immunol 148:2058–2061
36. Earronichi P, Macchia D, Piccinni M (1991) Allergen and bacterial antigen-specific T-cell clones established form atopic donors show a different profile of cytokine production. Proc Natl Acad Sci USA 88:4538–4542
37. Licinio J, Seibyl JP, Altemus M, et al. (1998) Elevated CSF levels of interleukin-2 in neuroleptic-free schizophrenic patients. Am J Psychiatry 150:1408–1410
38. Denicoff KD, Rubinow DR, Papa MZ, Simpson C, Seipp CA, Lotze MR, Change AE, Rosenstein D, Rosenberg SA (1987) The neuropsychiatric effects of treatment with interleukin-2 and lymphokine activated killer cells. Ann Intern Med 107:293–300
39. Rapaport MH, Lohr JB (1994) Serum interleukin-2 receptors in neuroleptic naïve schizophrenic subjects and in medicated schizophrenic subjects with and without tardive dyskinesia. Acta Psychiatr Scand 90:311–315
40. Rapaport MH, Caliguiri MP, Lohr JB (1997) An association between increased serum soluble interleukin-2 receptors and a disturbance in motor muscle force in schizophrenic patients. Prog Neuropsychopharmacol Biol Psychiatry 21(5):817–827
41. Schmidt RA, Zelaznik HN, Hawkins B, et al. (1979) Motor output variability: a theory for the accuracy of rapid motor acts. Psychol Rev 86:415–451
42. Schmidt RA (1982) Motor control and learning. University of Illinois Press, Champaign, IL
43. Bizzi E, Polit A, Morasso P (1976) Mechanisms underlying the achievement of final head position. J Neurophysiol 39:435–444

Correspondence: M. H. Rapaport, M.D., 8950 Villa La Jolla Dr. #2243, La Jolla, Ca. 92037, USA.

Measurement of neopterin, kynurenine and tryptophan in sera of schizophrenic patients

B. Sperner-Unterweger[1], **C. Miller**[1], **B. Holzner**[1], **B. Widner**[2], **W.W. Fleischhacker**[1], and **D. Fuchs**[2]

[1]Department of Biological Psychiatry, and [2]Institute of Medical Chemistry and Biochemistry, Innsbruck University Clinics, Innsbruck, Austria

Summary

Neopterin can be regarded as a sensitive marker of the activity of cell-mediated immunity. In chronic schizophrenic outpatients as well as in acute schizophrenic inpatients neopterin values were found within the normal range. Acutely ill schizophrenic patients had lowest neopterin levels at the day of admission, showing significantly decreased values when being compared to matched controls. Kynurenine data showed the same development as neopterin whereas tryptophan levels were rather unchanged.

These results support the hypothesis of an immune dysfunction in the acute state of schizophrenia.

Immunology being involved in the pathophysiology or in the etiology of schizophrenia has been addressed in numerous studies and although the results are controversial there are at least three hypotheses left which can be supported by the most consistent immunologic findings:

- an autoimmune process being involved in schizophrenic disorder (Knight, 1985; Ganguli, 1993, 1994)
- a viral infection or even slow virus infection being the cause of schizophrenic illness (Goodall, 1932; Menninger, 1928; Crow, 1984; DeLisi und Crow, 1986)
- an immunological dysregulation mainly refering to the T-cell system due to a yet unknown mechanism (Sperner-Unterweger, 1989; Hornberg, 1995; Rothermundt, 1996).

In the three outlined hypotheses the expected immunologic alterations are primarily refering to cell-mediated immunity (CMI). Assessing the activity of CMI, neopterin, a low molecular weight compound deriving biosynthetically from guanosine triphosphate can be regarded as a sensitive and validated marker (Fuchs, 1988). It is released in large quantities mainly by activated cells of the monocyte/macrophage lineage which have been stimulated by gamma-interferon (IFN-gamma) derived from activated T-cells (Huber, 1984; Werner, 1989; Wachter, 1992). Increased neopterin excretion in urine or in serum was demonstrated in diseases connected to cellular immune mechanisms (e.g., acute viral infections, graft vs. host disease, Aids) (Fuchs, 1988).

In two previous studies we measured urinary neopterin in chronic schizophrenic outpatients as well as in patients admitted to hospital due to acute schizophrenic psychosis (Sperner-Unterweger, 1989, 1992). For both patients groups neopterin levels were found within the normal range and the comparison between chronic schizophrenic patients and healthy controls did not show a statistically significant difference. Inpatients with acute symptomatology had lowest neopterin values at the day of admission, showing a statistically significant difference when being compared to healthy, sex- and age-matched controls ($p < 0.01$). After 7 days of treatment in hospital neopterin normalized and did not show any further differences in comparison to healthy probands. Evaluating the correlation between neopterin und psychopathology a positive correlation was found for patients suffering from chronic schizophrenia of the residual type whereas inpatients suffering from acute schizophrenia of the paranoid type showed a negative correlation (Sperner-Unterweger, 1989, 1992).

In a more recent study neopterin values, measured in sera of acutely ill schizophrenic patients, were found with the lowest concentrations at the day of admission followed by a significant increase at the end of treatment which consequently led to a significant difference in the comparison between schizophrenic patients and healthy controls ($p < 0.03$). The increase of neopterin observed in this study does not support the hypothesis of an CMI-activation in acute schizophrenia because neopterin values still remained inside the 95th percentile of age-matched controls. In accordance with our previous investigations (Sperner-Unterweger, 1989, 1992) and with the data published by Dunbar (1992) these results militate against the active involvement of viruses in the etiology of acute schizophrenia since neopterin can be regarded as a sensitive marker of viral infection such presenting with much higher neopterin values then those found in this study (Wachter, 1989). Similarly, the mainly normal neopterin values measured in these acutely ill schizophrenic patients stand in marked contrast to the elevations observed in autoimmune diseases (Wachter, 1989). It is important to note, however, that while CMI does not appear to be activated in these patients, the possibility remains that CMI might have been active earlier in the course of their illness, perhaps causing damage well before schizophrenia was clinically evident. Although in this study lowest neopterin values were found in the acute state of the illness followed by a slight increase occuring concomitantly with a continous stabilization of psychopathology it seems rather unlikely that antipsychotic medication could be responsible for this development because neither our group nor Dubar (1992) found an influence of medication on neopterin levels (Sperner-Unterweger, 1989, 1992).

Besides measuring neopterin, tryptophan and kynurenine concentrations were also evaluated in serum of the acutely ill schizophrenic patients. The kynurenine data of this investigation showed the same development as neopterin with lowest values on day 0 followed by a significant increase during treatment without presenting remarkable elevations which would be expected in the state of immune stimulation due to an acceleration of the kynurenine pathway metabolism (Heyes, 1992).

By interpreting the tryptophan data one has to keep in mind that tryptophan is substrate for two enzymic pathways: (1) tryptophan-5-hydroxylase converts tryptophan to 5-hydroxy-tryptophan to finally form the neurotransmitter 5-hydroxy-tryptamine (serotonin), (2) tryptophan pyrrolase and indoleamine (2,3)-dioxygenase (IDO) respectively, catalyse the degradation of tryptophan via the kynurenine pathway. The activity of tryptophan-5-hydroxylase as well as degradation of trytophan by IDO are both directly and indirectly associated with immune activation. IFN-gamma, released during states of cellular immune activation, will activate GTP-cyclohydrolase I in various cells, the key enzyme for the biosynthesis of pteridines. In most cells this will lead to increased formation of 5, 6, 7, 8-tetrahydrobiopterin and initiate the formation of serotonin if tryptophan 5-hydroxylase is present (Fuchs, 1991). By contrast, in human monocytes/macrophages large amounts of neopterin are formed instead of tetrahydrobiopterin (Fuchs et al., 1988; Werner, 1989). Our data presenting rather unchanged tryptophan levels during the observation periode do not support an acute state of CMI activation. Small fluctuations of tryptophan levels are probably due to changes in nutrition, assuming that some acutely ill schizophrenic patients do not eat regularly as long as they are outpatients; with admission to hospital they get regular meals containing sufficient amounts of essential aminoacids like tryptophan. The low kynurenine levels in patients at study entry would support the view of decreased endogenous levels of IFN-gamma.

Another study dealing with T-cell-subsets in schizophrenia showed highest values of total T-lymphocytes (CD3+) and T-helper cells (CD4+) in the acute state of schizophrenic psychosis, whereas remitted patients showed normal values. In contrast, NK-cells were found with lowest counts in acutely ill patients followed by normalization in remission of psychosis. Summarizing the results of these studies dealing with the cell mediated immune system in schizophrenia, the hypothesis of an immune dysfunction in the acute state of the illness is supported by low or subnormal neopterin, low kynurenine, low NK-cells, elevated CD3+ and CD4+ cells. These observations could be due to an inbalance of the TH-1 and TH-2 immunity (subclassification of T-helper-cells) (Parker, 1993). Decreased or low neopterin point to low or decreased production of IFN-gamma in schizophrenia. TH-1 and TH-2 derived cytokines negatively regulate the production of each other (Romagnani, 1994). This data would support the concept of a diminished TH-1 type immune response probably due to an activated TH-2 type immune response. The changed balance of these T-helper subtypes would be typical for a chronic infection. However, serological studies will be needet to confirm this assumptions.

References

Crow TJ (1984) A re-evaluation of the ciral hypothesis: is psychosis the result of retroviral integration at a side close to the cerebral dominance gene? Br J Psychiatry 145:243–253

DeLisi LE, Crow TJ (1986) Is schizophrenia a viral or immunologic disorder? Psychiat Clin North Am 9:115–132

Dunbar PR, Hill J, Neale TH, Mellsop GW (1992) Neotperin measurement provides evidence of altered cell-mediated immunity in patients with depression, but not with schizophrenia. Psychol Med 22:1051–1057

Fuchs D, Hausen A, Reibnegger G, Werner ER, Dierich MP, Wachter H (1988) Neopterin as a marker for activated cell-mediated immunity: application in HIV infection. Immunol Today 9:150–155

Fuchs D, Weiss G, Werner-Felmayer G, Wachter H (1991) Cytokine-induced increase in liver serotonin. Immunol Lett 28:259

Ganguli R, Brar HS, Chengappa KMR, Yang ZW, Nimgaonkar VL, Rabin BS (1993) Autoimmunity in schizophrenia: a review of recent findings. Ann Med 25:489–496

Ganguli R, Brar HS, Rabin BS (1994) Immune abnormalities in schizophrenia: evidence for the autoimmune hypothesis. Harvard Rev Psychiatr 2:70–83

Goodall E (1932) The existing couse of certain states, at present classified under "schizophrenia" by psychiatrists may be infection. J Ment Sci 78:746

Heyes MP, Saito K, Crowley JS, Davis LE, Demitrack MA, Der M, Dilling LA, Elia J, Kruesi MJP, Lckner A, Larsen SA, Lee K, Leonard HL, Markey P, Martin A, Milstein S, Mouradian MM, Pranzatelli MR, Quearry BJ, Salazar A, Smith M, Strauss SE, Sunderland T, Swedo SW, Tourtellotte WW (1992) Quinolinic acid and kynurenine pathway metabolism in inflammatory and non-inflammatory neurological disease. Brain 115:1249–1273

Hornberg M, Arolt V, Wilke I, Kruse A, Kirchner H (1995) Production of interferons and lymphokines in leukocyte cultures of patients with schizophrenia. Schizophr Res 15:237–242

Huber C, Batchelor JR, Fuchs D, Hausen A, Lang A, Niederwieser D, Reibnegger G, Swetly P, Troppmair J, Wachter H (1984) Immune response-associated production of neopterin: release from macrophages primarily under control of interferon gamma. J Exp Med 160:310–316

Knight J (1985) Possible autoimmune mechanisms in schizophrenia. Integr Psychiat 3:134
Menninger K (1928) The schizophrenic syndrome as a product of acute infectious disease. Arch Neurol Psychiatry 20:464
Parker DC (1993) T cell-dependent B cell activation. Ann Rev Immunol 46:111–147
Romagnani S, Del Prete G, Manetti R, Ravina A, Annunziato F, Delarli M, Mazzetti M, Piccinni MP, D, Elios MM, Parronchi P (1994) Role of TH1/TH2 cytokines in HIV infection. Immunol Rev 140:73–92
Rothermundt M, Arolt V, Weitzsch C, Eckhof D, Kirchner H (1996) Production of cytokines in acute schizophrenic psychosis. Biol Psychiatry 40:1294–1297
Sperner-Unterweger B, Barnas C, Fleischhacker WW, Fuchs D, Meise U, Reignegger C, Wachter H (1989) Is schizophrenia linked to alteration in celloular immunity? Schizophr Res 2:417–421
Sperner-Unterweger B, Barnas C, Fuchs D, Kemmler G, Wachter H, Hinterhuber H, Fleischhacker WW (1992) Neopterin production in acute schizophrenic patients: an indicator of alterations of cellmediated immunity. Psychiat Res 42:121–128
Wachter H, Fuchs D, Hausen A, Reibnegger G, Werner ER (1989) Neopterin as marker for activation of cellular immunity: immunologic basis and clinical application. Adv Clin Chem 27:81–141
Wachter H, Fuchs D, Hausen A, Reibnegger G, Weiss G, Werner ER, Werner-Felmayer G (1992) In: Neopterin: biochemistry, methods and clinical application. de Gruyter, Berlin New York
Werner ER, Werner-Felmayer G, Fuchs D, Hausen A, Reibnegger G, Wachter H (1989) Parallel induction of tetrahydrobiopterin biosynthesis and indoleamine 2,3-dioxygenase activity in human cells and cell lines by interferon-gamma. Biochem J 262:861–866

Correspondence: Dr. B. Sperner-Unterweger, Department of Biological Psychiatry, Innsbruck University Clinics, Anichstrasse 35, A-6020 Innsbruck, Austria

Levels of soluble adhesion molecules in schizophrenia: relation to psychopathology

M.J. Schwarz, M. Riedel, M. Ackenheil, and **N. Müller**

Psychiatric Hospital, Ludwig-Maximilian University, Munich,
Federal Republic of Germany

Summary

The intercellular adhesion molecule-1 (sICAM-1) and the vascular adhesion molecule-1 (VCAM-1) are involved in the penetration of lymphocytes through the blood-brain-barrier (BBB). Since a BBB disturbance has been observed in some schizophrenic patients, but also a lack of the CNS/peripheral immune system has been hypothesized to be involved in schizophrenia, we measured the serum levels of soluble ICAM-1 (sICAM-1) and soluble VCAM-1 (sVCAM-1) in unmedicated schizophrenic patients and the relationship to psychopathology. The results show lower levels of sICAM-1 in schizophrenics, but no difference between schizophrenics and controls regarding the sVCAM-1 levels. Patients with more pronounced negative symptoms showed higher levels of sICAM-1 and lower levels of sVCAM-1. These results point out that the ICAM-1 related adhesion system is involved in schizophrenia, while the VCAM-1 system plays a minor role. The regulatory process of antigen recognition and cellular immune activation and/or the regulation of the BBB is impaired at least in a subgroup of schizophrenic patients.

Introduction

The intracellular adhesion molecule-1 (ICAM-1) and the vascular adhesion molecule-1 (VCAM-1) are expressed on blood cells and tissue. Both molecules are involved in the adhesion of lymphocytes and the penetration of immune cells from the blood into target organs. Adhesion molecules are activated by proinflammatory cytokines such as interleukin-1 (IL-1) or IL-2 (De Vries et al., 1994), which are discussed to be involved in the pathogenesis of schizophrenia (Licinio et al., 1993; Katila et al., 1995; McAllister et al., 1995; Sirota et al., 1995).

A multiple receptor – ligand system mediates adhesion. A great variety of cell adhesion molecules belongs to this system (Springer,

1990). The cell adhesion system has three functions: (**1**) accumulation in the vessels of the target organ by adhesion, (**2**) diapedesis through the endothelium of certain lymphocyte subpopulations, and (**3**) traffic into the parenchyma (Raine et al., 1990; Hampel et al., 1996). The ligands of VCAM-1 and ICAM-1, the very late antigen-4 (VLA-4) and the lymphocyte function-related antigen-1 (LFA-1) are expressed on leukocytes and on endothelium. In the CNS, VCAM-1 and ICAM-1 are expressed on astrocytes and microglial cells, and also on neurons (Raine et al., 1990; Fabry et al., 1992; Héry et al., 1995; Hampel et al., 1996).

VCAM-1 and ICAM-1 are part of the regulatory system of the blood-brain barrier (BBB). Soluble ICAM-1 (sICAM-1) is a reliable marker of BBB impairment (Rieckmann et al., 1993; Schwarz et al., 1998a). Serum levels of sICAM-1 are related to the expression of ICAM-1 on the cell surface, because sICAM-1 is shed from the cell surface (Rothlein et al., 1991). Moreover, since one of the major sources of sICAM-1 are activated macrophages (Krenn et al., 1997), the key cells for antigen presentation, and activation of the immune answer, the level of sICAM-1 in the blood is a marker for the activation of the immune system. On the other hand, ICAM-1 expression on endothelial cells mediates the lymphocyte traffic into CNS parenchyma, and thus sICAM-1 levels are also related to a functioning BBB.

Since a disturbance of the BBB is found in about 20%–30% of schizophrenic patients and especially schizophrenic negative symptoms are related to the albumin and immunoglobulin G levels of the CSF (Müller and Ackenheil, 1995), we hypothesized that the levels of sICAM-1 and of sVCAM-1 are also altered in schizophrenia and related to the schizophrenic symptomatology, especially to the schizophrenic negative symptoms.

Material and methods

Serum samples were obtained from 36 unmedicated schizophrenics (16 females, 20 males; mean age 33 ± 10 years, range from 18 to 63 years) with a mean duration of illness of 70 ± 90 months. All patients fulfilld the DSM III-R criteria for schizophrenia. Sixteen of the patients were having their first episode of schizophrenia and had never been medicated with neuroleptics. The remaining 20 patients had a wash-out period for at least four months before the study. Psychopathology was monitored using the PANSS score before treatment at the time of blood sampling.

Informed consent was obtained from all subjects participating in the study.

Control serum samples were taken from 38 age and sex matched healthy subjects (15 females, 23 males, mean age 28 ± 6 years). Individuals suffering from infectious or autoimmune disorders or alcohol abuse were excluded from the study; nicotine abuse was noted. To exclude any additional disease, blood count, blood sedimentation rate, and C-reactive protein were measured in all serum samples. In 28 of the patients and in 30 of the control subjects, soluble

VCAM was also measured (mean age of patients: 32 ± 10; mean age of controls: 29 ± 7 years).

The soluble proteins sICAM-1 and sVCAM were estimated with the following commercially available ELISA kits: ICAM-1 (Cellfree® sICAM-1 ELISA, Endogen, USA) and human sVCAM-1 (R&D Systems, Minneapolis, USA). The laboratory staff was not informed about the subject and diagnosis of the serum probes.

Statistics

For statistical evaluation we used Pearson's correlation test, t test for independent samples, t test for paired samples, one-way ANOVA, and Mann-Whitney U–Wilcoxon rank sum W test. The program SPSS 6.1.3 (SPSS Inc.) was used for statistical analysis.

Results

Serum concentrations of sICAM-1 were significantly lower in unmedicated patients than in the healthy control subjects (patients: 247 ± 94 ng/ml; controls: 321 × 72 ng/ml; independent samples' t test: t = −3.76; p = .000) (see Fig. 1).

In contrast, serum levels of soluble VCAM-1 did not differ between the groups (patients: 506 ± 120 ng/ml; controls: 559 ± 139 ng/ml) (see Fig. 2).

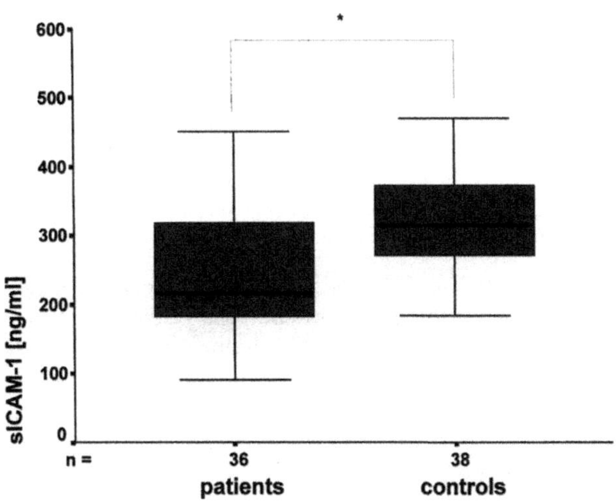

Fig. 1. Serum levels of soluble ICAM-1 are significantly lower in 36 unmedicated schizophrenic patients than in 38 healthy controls

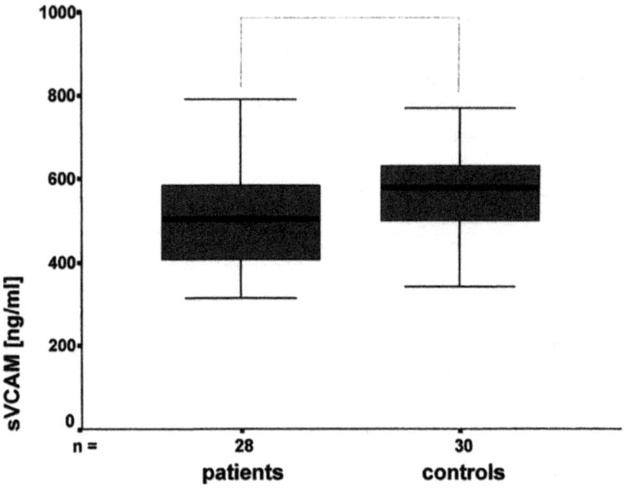

Fig. 2. Serum levels of soluble VCAM-1 in 28 unmedicated schizophrenic patients and 30 healthy controls

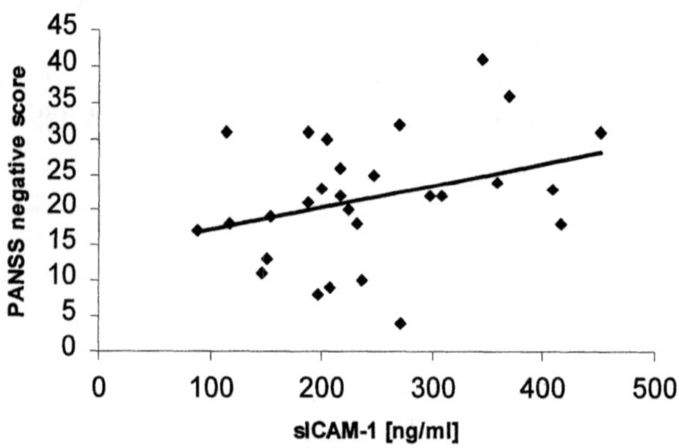

Fig. 3. Significant positive relation between serum levels of soluble ICAM-1 and schizophrenic negative symptomatology rated by PANSS negative score

Gender, age, duration of illness, nicotine abuse, or the occurance of a first episode of schizophrenia, had no influence on sICAM-1 or sVCAM-1 levels.

High concentrations of sICAM-1 in unmedicated patients were directly related to high values for the PANSS-negative score and the PANSS-total score (negative score: r = .39, p = .019; total: r = .38, p = .024) (see Fig. 3).

Concentrations of sVCAM, however, showed an inverse relation to psychopathology: high levels of sVCAM-1 were related to low PANSS-

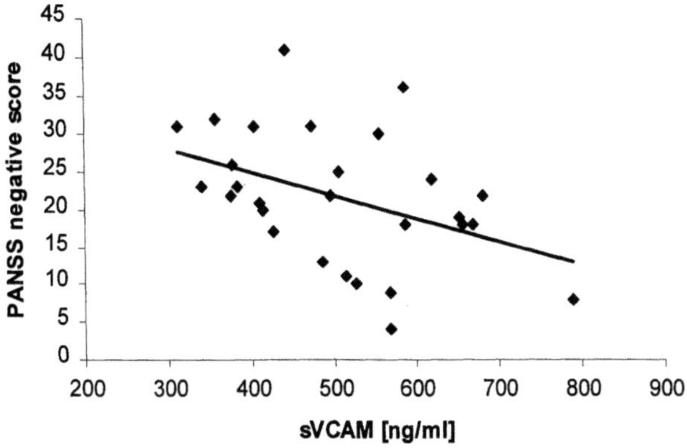

Fig. 4. Significant negative relation between serum levels of soluble VCAM-1 and schizophrenic negative symptomatology rated by PANSS negative score

negative scores and to low PANSS-total scores (negative: $r = -.43$, $p = .021$; total: $r = -.42$, $p = .026$) (see Fig. 4).

Discussion

Two major findings have to be highlighted: the decreased levels of sICAM-1 in schizophrenics, and the relation between soluble adhesion molecules and schizophrenic psychopathology.

Psychopathology

The close relation between sICAM-1 levels and schizophrenic negative symptoms agrees with earlier findings (Müller and Ackenheil, 1995). Although no direct interrelation between BBB parameters and schizophrenic negative symptoms has been observed before, high CSF-albumin and -IgG levels, which correlated with more pronounced negative symptoms, are often found in patients with a BBB disturbance. Since sICAM-1 levels are related to a BBB disturbance (Schwarz et al., 1998a), higher scores for negative symptoms would be expected in patients with higher sICAM-1 levels. Accordingly, the higher PANSS total score in patients with high sICAM-1 levels seems to be mainly due to the higher PANSS negative score. However, the biological meaning of decreased sICAM-1 levels and the possible mechanisms behind this relation between schizophrenic negative symptoms and sICAM-1 levels are unclear.

Definitive conclusions about sVCAM-1 levels cannot be drawn either. It seems that sVCAM-1 plays not a pivotal role in schizophrenia, because its serum concentration is not altered. However, the inverse relation between sVCAM-1 levels and the schizophrenic negative symptoms needs to be explained. It indicates that high sVCAM-1 concentrations are associated with low schizophrenic symptoms and vice versa. This finding suggests two conclusions. First, sVCAM-1 is also part of the regulatory system of the BBB, and an impairment of CNS-blood communication mediated by sVCAM-1 may be associated with more marked disease symptoms. Second, the VCAM-1/VLA-4$^+$ system acts as a co-stimulatory signal in the immune activation and antigen recognition (Burkley et al., 1991). Low sVCAM-1 levels may indicate a lack of immune activation, which accompanies more pronounced schizophrenic symptoms.

sCAM-levels

While increased levels of sICAM-1 occur in inflammatory and autoimmune disorders, including those of the CNS (Rieckmann et al., 1993; Sharief et al., 1993), decreased levels of sICAM-1 have not been reported in the literature. They have, however, been observed in schizophrenic patients (G. Wieselmann, personal communication). Since our schizophrenic patients have not been treated with antipsychotic medication, low levels of sICAM-1 cannot be an effect of treatment. Other aspects of sICAM-1 levels in schizophrenic patients have been investigated in our group: Relatively increased sICAM-1 serum concentrations are found in those schizophrenis patients who also show other signs of immune activation, e.g., increased levels of heat-shock protein 60 (Schwarz et al., 1998b). In CSF higher levels of sICAM-1 levels were found to be related to an impairment of the BBB (Schwarz et al., 1998a).

Since ICAM-1 has two different functions, a disturbance of the ICAM-1 system may have different implications for schizophrenia. These two functions are (1) a costimulatory signal in immune activation and (2) the facilitation of adhesion and penetration through the BBB.

In the first function ICAM-1 acts as an important costimulatory molecule for lymphocyte activation (Van Seventer et al., 1990; Kuhlmann et al., 1991). When ICAM-1 expression is decreased, the peripheral immune system may not be activated accordingly. The concept of an impaired stimulatory signal in schizophrenia fits well with the hypothesis that antigen presentation/recognition is impaired in acutely ill schizophrenics. This hypothesis is supported by the following observations: schizophrenic patients show a reduced lymphocyte stimulation to specific antigens (Müller et al., 1991), a reduced production of interferon-γ in lymphocyte cultures (Rothermund et al., 1996), and a clearly reduced immune answer to hepatitis B vaccination (Russo et al., 1994) compared with controls.

In the second function, low levels of sICAM-1 may indicate impaired communication between the peripheral immune system and the CNS. It can be speculated that decreased sICAM-1 levels reflect a disturbance in the T-cell system, which penetrate the BBB and patroill within the brain. Since an increase of activating cytokines such as IL-2 or IL-6 may be involved in the pathogenesis of schizophrenic psychoses, an activation of the peripheral immune system including an increase of ICAM-1 expression in parallel to an increase of activating cytokines in the CSF would be expected. However, impaired communication between the CNS and the peripheral immune system may also result in inadequate activation of the peripheral immune system, including insufficient activation of the adhesion system (Marker et al., 1995). CNS viral infections have shown that immune activation in the CNS and increased communication between the CNS and peripheral immune systems predict a better outcome than impaired activation, that can be associated with poor outcome (Lewandowski et al., 1994).

Several studies have reported signs of an autoimmune process in the peripheral immune system of one-third of schizophrenics, often in chronic cases or cases with a poor therapeutic outcome (McAllister et al., 1989; Kilidirias et al., 1992; Müller and Ackenheil, 1995). Antipsychotic treatment seems to contribute to an activation of the immune system and to antibody production (Özek et al., 1971; Gallien et al., 1977; Zarrabin et al., 1979; Müller et al., 1997). During antipsychotic treatment increases the expression of the cellular surface molecules VLA-4 and LFA-1 on $CD4^+$ cells (Müller et al., 1999) and it has been observed in another study that the levels of sICAM-1 increase slightly, but not significantly during antipsychotic treatment (Schwarz et al., 1999). Further studies have to evaluate whether especially high sICAM-1 levels correspond not only with an impaired BBB but also with longer or more intense antipsychotic medication. However, the physiological mechanism of sICAM-1 is not yet fully understood. Although the serum level of sICAM-1 is related to the cellular expression of ICAM-1, sICAM-1 acts as a functional antagonist of ICAM-1 that inhibits lymphocyte attachment to cerebral endothelial cells (Rieckmann et al., 1995).

Moreover, the interelation between the regulatory mechanism in the ICAM-1/LFA-1 system and the VCAM-1/VLA-4 system is only partly understood (Oppenheimer-Marks et al., 1991).

Our findings suggest that sICAM-1 may be strongly involved in schizophrenia, whereas sVCAM-1 plays a minor role. Moreover, the regulatory process of antigen recognition and immune activation of T-cells and/or the regulation of the BBB may be impaired in schizophrenia.

References

Burkly LC, Jakubowski A, Newman BM, Rosa MD, Chi-Rosso G, Lobb RR (1991) Signaling by vascular cell adhesion molecule-1 (VCAM-1) through VLA-4 promotes CD3-dependent T cell proliferation. Eur J Immunol 21:2871–2875

De Vries HE, Moor ACE, Blom-Roosemalen MCM, de Broer AG, Breimer DD, van Berkel TJC, Kuiper J (1994) Lymphocyte adhesion to brain capillary endothelial cells in vitro. J Neuroimmunol 52:1–8

Fabry Z, Waldschmidt MM, Hendrickson D, Keiner J, Love-Homann L, Takei F, Hart MN (1992) Adhesion molecules on murine brain microvascular endothelial cells: expression and regulation of ICAM-1 and Lgp 55. J Neuroimmunol 36:1–11

Gallien M, Schnetzler JP, Morin J (1977) Antinuclear antibodies and lupus cells in 600 hospitalized phenothiazine treated patients. Ann Med Psychol Med 1:237–248

Hampel H, Schwarz M, Kötter HU, Schneider C, Müller N (1996) Cell adhesion molecules in the central nervous system. Drug News Perspect 9:69–81

Héry Ch, Sébire G, Peudenier S, Tardieu M (1995) Adhesion to human neurons and astrocytes of monocytes: the role of interaction of CR3 and ICAM-1 and modulating by cytokines. J Neuroimmunol 57:101–109

Katila H, Appelberg B, Hurme M, Rimon R (1995) Plasma levels of Interleukin-1 beta and Interleukin-6 in schizophrenia, other psychoses, and affective disorders. Schizophr Res 12:29–34

Kilidireas K, Latov N, Strauss DH, Gorig AD, Hashim GA, Gorman JM, Sadiq SA (1992) Antibodies to the human 60 kDa heat-shock protein in patients with schizophrenia. Lancet 340:569–572

Krenn V, Schedel J, Döring A, Huppertz HI, Gohlke F, Tony HP, Vollmers HP, Müller-Hermelink HK (1997) Endothelial cells are the major source of sICAM-1 in rheumatoid synovial tissue. Rheumatol Int 17:17–27

Kuhlmann P, Moy VT, Lollo BA, Brian AA (1991) The accessory function of murine intercellular adhesion molecule-1 in T-lymphocyte activation: contribution of adhesion and co-activation. J Immunol 146:1773–1782

Lewandowski G, Hobbs MV, Blomom FE (1994) Alteration of intracerebral cytokine production in mice infected with herpes simplex virus types I and 2. J Neuroimmunol 55:23–34

Licinio J, Seibyl JP, Altemus M, Charney DS, Krystal JH (1993) Elevated levels of Interleukin-2 in neuroleptic-free schizophrenics. Am J Psychiatry 150:1408–1410

Marker O, Scheynius A, Christensen JP, Thomsen AR (1995) Virus-activated T cells regulate expression of adhesion molecules on endothelial cells in sites of infection. J Neuroimmunol 62:35–42

McAllister CG, Rapaport MH, Pickar D, Podruchny TA, Christison G, Alphs LD, Paul SM (1989) Increased number of CD5+ B-lymphocytes in schizophrenic patients. Arch Gen Psychiatry 46:890–894

McAllister CG, van Kammen DP, Rehn TJ, Miller AL, Gurklis J, Kelley ME, Yao J, Peters JL (1995) Increases in CSF levels of interleukin-2 in schizophrenia: effects of recurrence of psychosis and medication status. Am J Psychiatry 152:1291–1297

Müller N, Ackenheil M (1995) Immunoglobulin and albumin contents of cerebrospinal fluid in schizophrenic patients: the relationship to negative symptomatology. Schizophr Res 14:223–228

Müller N, Ackenheil M (1998) Psychoneuroimmunology and the cytokine-network in the CNS: implications of psychiatric disorders. Prog Neuropsychopharmacol Biol Psychiatry 22:1–33

Müller N, Ackenheil M, Hofschuster E, Mempel W, Eckstein R (1991) Cellular immunity in schizophrenic patients before and during neuroleptic therapy. Psychiatry Res 37:147–160

Müller N, Riedel M, Schwarz MJ, Gruber R, Ackenheil M (1997) Immunomodulatory effects of neuroleptics to the cytokine system and the cellular immune system in schizophrenia. In: Wieselmann G (ed) Current update in psychoimmunology. Springer, Wien New York, pp 57–67

Müller N, Hadjamu M, Riedel M, Schwarz MJ, Ackenheil M, Gruber R (1999) Adhesion-molecule receptor expression on T-cells and the role of the blood-brain-barrier in schizophrenia. Am J Psychiatry 156:634–636

Özek M, Toreci K, Akkok M, Guvener Z (1971) Influence of neuroleptic therapy on antibody formation. Psychopharmacologia (Berl) 21:401–412

Oppenheimer-Marks N, Davis LS, Bogue DT, Ramberg J, Lipsky PE (1991) Differential utilisation of ICAM-1 and VCAM-1 during the adhesion and transendothelial migration of human T-lymphocytes. J Immunol 147:2913–2921

Raine CS, Canella B, Duijvestijn AM, Cross AH (1990) Homing to central nervous vasculature by antigen-specific lymphocytes. II. Lymphocyte/endothelial cell adhesion during the initial stages of autoimmune demyelination. Lab Invest 63:476–479

Rieckmann P, Nunke K, Burchhardt M, Albrecht M, Wiltfang J (1993) Soluble intercellular adhesion molecule-1 in cerebrospinal fluid: an indicator for the inflammatory impairment of the blood-cerebrospinal fluid barrier. J Neuroimmunol 47:133–140

Rieckmann P, Michel U, Albrecht M, Bruck W, Wockel L, Felgenhauer K (1995) Soluble forms of intercellular adhesion molecule-1 (ICAM 1) block lymphocyte attachment to cerebral endothelial cells. J Neuroimmunol 60:9–15

Rothermund M, Arolt V, Weitzsch C, Eckhoff D, Kirchner H (1996) Production of Interferon gamma is reduced in acute schizophrenia. Biol Psychiatry 40:1294–1296

Rothlein R, Mainolfi EA, Czajkowski M, Marlin SD (1991) A form of circulating ICAM-1 in human serum. J Immunol 147:3788–3793

Russo R, Ciminale M, Ditommaso S, Siliquini R, Zotti C, Ruggenini AM (1994) Hepatitis B vaccination in psychiatric patients. Lancet 343:394

Schwarz MJ (1998) Die Bedeutung von Autoantikörpern gegen 60 kD und 70 kD Hitzeschock-Proteine für die Pathogenese schizophrener Erkrankungen. Thesis, University of Munich

Schwarz MJ, Ackenheil M, Riedel M, Müller N (1998a) Blood-CSF barrier impairment as indicator for an immune process in schizophrenia. Neurosci Lett 254:1–3

Schwarz MJ, Riedel M, Ackenheil M, Müller N (1998b) Autoantibodies against heat shock-protein 60 in schizophrenia. Eur Arch Psychiatry Clin Neurosci 248:282–288

Schwarz MJ, Ackenheil M, Riedel M, Müller N (1999) Soluble intercellular adhesion molecule-1 (ICAM-1) receptor levels are decreased in unmedicated and medicated schizophrenic patients. Biol Psychiatry (submitted)

Sharief MK, Noori MA, Ciardi M, Cirelli A, Thompson EJ (1993) Increased levels of circulating ICAM-1 in serum and cerebrospinal fluid of patients with active multiple sclerosis. Correlation with TNF-alpha and blood-brain barrier damage. J Neuroimmunol 43:15–21

Sirota P, Schild K, Elizur A, Djaldetti M, Fishman P (1995) Increased Interleukin-1 and Interleukin-3 like activity in schizophrenic patients. Prog Neuropsychopharmacol Biol Psychiatry 19:75–83

Springer TA (1990) Adhesion receptors of the immune system. Nature 346:425–434

Van Seventer GA, Shimizu Y, Horgan KJ, Shaw S (1990) The LFA-1 ligand ICAM-1 provides an important costimulatory signal for T-cell receptor mediated activation of resting T-cells. J Immunol 144:4579

Zarrabi MH, Zucker S, Miller F, Derman RM, Romano GS, Hartnett JA, Varma AO (1979) Immunologic and coagulation disorders in chlorpromazine-treated patients. Ann Int Med 91:194–199

Correspondence: Dr. N. Müller, Priv.-Doz., Dipl.-Psych, Psychiatric Hospital, Ludwig Maximilian-University, Nußbaumstrasse 7, D-80336 München, Federal Republic of Germany

The role of interferon-alpha in the regulation of sleep

J. Born[1] and **E. Späth-Schwalbe**[2]

[1]Department of Clinical Neuroendocrinology, University of Luebeck, and Department of Physiological Psychology, University of Bamberg, and [2]Department of Internal Medicine, Humboldt University, Charité, Berlin, Federal Republic of Germany

Summary

Interferon-alpha (IFN-alpha) exerts strong antiviral effects and was among the first cytokines suspected to contribute to alterations of sleep during infectious challenge. A short review of the relevant literature is provided. Administration of IFN-alpha in rabbits increased body temperature and, in parallel, slow wave sleep (SWS). These findings were taken to suggest a sleep promoting effect of the cytokine. However, experiments in monkeys, although indicating increased behavioral signs of fatigue, did not reveal changes towards enhanced sleep after IFN-alpha. In cancer patients treated with high doses of IFN-alpha enhanced fatigue is commonly observed. We conducted a first study in healthy young men, evaluating the effects of IFN-alpha given at a fairly low dose (1000 and 10,000 IU/kg body weight, subcutaneously) on nocturnal sleep. The cytokine was administered subcutaneously at 19.00 h prior to sleep. It distinctly reduced SWS during the early night and, at the high dose, REM sleep across the entire night. Effects on sleep were accompanied by increased feelings of tiredness, increased pituitary-adrenal secretory activity, increased growth hormone secretion and increased IL-6 plasma concentrations. Body temperature remained almost unchanged. These results support an acute disrupting effect of IFN-alpha on nocturnal sleep in humans. Similar disruptive effects have been observed in previous studies after administration of IL-6 in healthy humans. Together, the data support the view that enhanced release of proinflammatory cytokines in humans acutely impairs sleep. Given that the pattern of changes after IFN-alpha and also after IL-6 is in some of its essential features is reminiscent of the symptoms seen in depressed patients, a contribution of host defense mechanisms to this disease and associated sleep disruption should be taken into consideration.

Introduction

Interferons (IFNs) represent a class of cytokines which is most critical in the regulation of the antiviral defense (DeMaeyer and DeMaeyer-Guignard, 1994, for a review of immunological properties). IFN-alpha like IFN-β, belongs to the type I IFNs which act via the same cell-surface receptor. Probably all cells of the organism including those of the central nervous system, are capable to produce IFN-alpha, although not spontaneously. Viruses are the most efficient natural inducer of IFN-alpha, but other infectious agents can also induce the production of this cytokine. IFN-alpha differes significantly from type II IFN-gamma which is produced mainly by lymphocytes and natural killer cells, binds to separate receptors and displays less intiviral activity than IFN-alpha. Released upon viral infection, IFN-alpha is most effective in prohibiting the proliferation of the infected cells and in inducing the expression of class I MHC molecules thereby facilitating the recognition of the endogenous target antigens. In vitro, IFN-alpha has been shown to facilitate the production of a number of other cytokines including IL-1β and IL-6.

IFN-alpha was one of the first cytokines suspected to induce feelings of sleepiness and sleep associated with the response to infectious agents. This view was stimulated also by clinical observations that in some infections known to be caused by viruses changes in the structure of sleep and of the sleep-wake rhythm belong to the most prominent symptoms. Encephalitis lethargica which is presumed to be a viral central nervous infection, was epidemic at the beginning of this century and in more than 80% of the patients was associated with severe alterations of sleep (von Economo, 1930). Mononucleosis resulting from infection with Epstein Barr Virus in a subgroup of patients is associated with increased total sleep time and shortened sleep latency (Guillminaut and Mondini, 1986). Distinct changes in sleep have been revealed also in patients infected with the immunodeficiency virus (HIV), even during the asymptomatic period (Norman et al., 1990, 1992; Kubicki et al., 1988). Correspondingly, the chronic fatigue syndrom characterized by daytime tiredness and unrefreshing sleep, has been supposed to be induced by viral infection (Moldofsky, 1995). Together, these observations tempted speculations that the marked alterations of sleep representing major symptoms of psychiatric disorders such as depression and schizophrenia could likewise point to an underlying infectious process.

However, the observation of changes in sleep architecture in conjunction with infectious challenge does not allow for any conclusion as to whether or not the sleep alteration reflect a specific regulatory aspect of host defense. To resolve this issue, studies were undertaken in animals and, more recently, also in humans which aimed to determine the influence of specific cytokines on regular sleep. Some of these studies have been devoted to the effects of IFN-alpha.

Effects of IFN-alpha on sleep in animals

Changes in sleep after administration of IFN-alpha were investigated in rabbits, rats and monkeys. Krueger and coworkers (1987) treated two groups of male New Zealand white rabbits intravenously with, respectively, 15×10^6 U human recombinant IFN-alpha2 and intracerebroventricularly with $1–2 \times 10^6$ U IFN-alpha2. On the corresponding control conditions, saline (0.5 ml) and artificial cerebrospinal fluid (0.3 ml) was administered. Substances were administered between 8.00 and 10.00 h. (Although domestic rabbits usually display increased activity during the day, sleep occupied up to 50% of the daylight hours in these animals). Figure 1 summarizes the effects of IFN-alpha after the

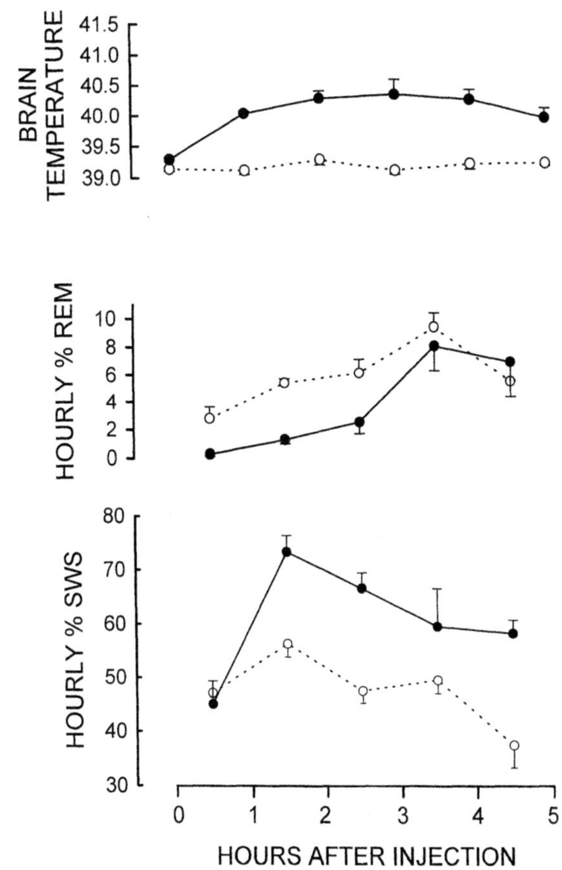

Fig. 1. Effects of intravenous administration of IFN-alpha2 (15×10^6 U, solid lines) versus saline (dashed lines) on brain temperature, percentages of REM sleep and SWS in rabbits (n = 6). Means (+/– SEM) are indicated. IFN-iduced fever and excess of SWS was first evident during the second hour post-infusion, then persisted for the rest of the assay period. The reduction in REM sleep following IFN-alpha2 was signifcant during the first 3 hours post-injection (from Krueger et al., 1987)

intravenous administration. Brain temperature increased within 3 hours by 1.1°C on average, and the cytokine concurrently enhanced the percentage of time spent in SWS. The time course of the effect on SWS paralleled that of the fever response: While during the first hour after IFN-alpha administration, time in SWS was similar to that obtained in the control condition, SWS distinctly increased during the second hour post-infusion. This increase persisted during the remaining 3 hours of the recording epoch. An inhibiting effect of IFN-alpha-2 on REM sleep appeared to be less pronounced, although this effect revealed to be highly signifcant during the first 3 hours post-infusion. Even stronger increases in brain temperature and SWS were obtained after intracerebroventricular administration of the cytokine. But contrasting with the effects of intravenous IFN-alpha, the intracerebroventricular route of administration did not induce a transient suppression of REM sleep. The cyclic structure of sleep and waking seemed to be unaffected by the cytokine.

Considering that the intravenous dose of IFN-alpha required to elicit SWS effects was only 10–15 times the effective intracerebroventricular dose, Krueger et al. (1987) suggested that the cytokine elicited its sleep promoting effect most likely via a messenger system with direct and easy access to those brain structures controlling sleep and SWS. Even IFN-alpha itself can enter the brain after systemic administration (Habif et al., 1975; Smith et al., 1985), and receptors binding IFN-alpha have been identified in central nervous tissues (Janicki, 1992). Since, the suppression of REM sleep by IFN-alpha in the rabbits was seen only with the intravenous route of administration, this effect in contrast with that on SWS, appears to reflect a primary systemic action, possibly by stimulating pituitary-adrenal release of adrenocorticotropin (ACTH) and corticosteroids (Späth-Schwalbe et al., 1989). Administration of these hormones has been shown to inhibit REM sleep, although in one study in rabbits the effect turned out less consistent (Kawakami et al., 1965; Born et al., 1989).

The view of an enhancing effect of IFN-alpha on sleep and SWS received some support from subsequent experiments in rats (DeSarro et al., 1989). In those experiments, IFN-alpha microinjected into the locus ceruleus induced a transient enhancement of sleep and EEG slow wave activity. However, promoting effects on sleep and SWS have been likewise observed following administration of IL-1β in rabbits and rats. Considering that IFN-alpha can induce the production of IL-1β, the somnogenic effect of IFN-alpha could prove to be mediated by IL-1β (Krueger and Majde, 1994).

Experiments in monkeys yielded results somewhat diverging from those in rabbits and rats. Reite et al. (1987) intravenously injected 20 $\times 10^6$ U/m^2 IFN-alpha in pigtail monkeys at approximately 8.30 h in the morning following a baseline night of sleep. The effects on sleep were recorded in the two succeeding nights. A total of 13 IFN injections was studied. The animals exhibited increased evidence of behavioral sleepiness on the days IFN-alpha was administered. However, sleep architecture appeared to be unaffected by the cytokine, except for a distinct decrease in REM sleep latency. The effect was present on both nights following injection of IFN-alpha. Notably, a decrease in REM latency in conjunction with subjective tiredness is frequently observed also in depressed patients suffering from sleep disturbances.

Effects of IFN-alpha on sleep in humans

In clinical trials in cancer patients treated for extended periods of time with high doses of IFN-alpha observations have been made of an increased sleepiness and lassitude in these patients (e.g., Smedley et al., 1983). However, somnopolygraphic recordings of sleep were not conducted in those studies. Also, the doses used for therapeutic purposes induced blood IFN concentrations which presumably are hardly comparable with the increases accompanying normal viral infection. Here, we will therefore concentrate on a first study which we conducted in young healthy men to determine the influence of a single administration of rather low doses of IFN-alpha on nocturnal sleep (Späth-Schwalbe et al., 1999). This study is the second in an ongoing series of studies initiated in our laboratory which aims to differentiate effects of proinflammatory and inhibitory cytokines on human sleep. The first study (Späth-Schwalbe et al., 1998), investigated somnogenic effects of IL-6, and has been discussed in greater detail in a previous volume of this series (Born and Späth-Schwalbe, 1997).

The effects of human recombinant IFN-alpha subcutaneously administered at doses of 1000 and 10,000 IU/kg body weight, were evaluated in a placebo controlled double-blind study in 18 healthy men aged between 21 and 31 years. The subjects were screened by medical history, a physical examination, laboratory investigations and electrocardiogram to exclude any acute and chronic illness. They had normal sleep-wake rhythms and had not participated in night work for at least 4 weeks prior to the experimental nights. The study was designed according to a within-subject comparison. After a night of adjustment to the laboratory, each man participated in 3 experimental nights separated by at least 10 days.

An experimental night lasted between 18.00 and 8.00 h. Throughout this period the subject stayed in bed in a supine position. Placebo and the two doses of IFN-alpha were injected at 19.00 h before lights were turned off at 23.00 h. Sleep was allowed until 7.00 h the next morning, when the man was awakened. At 21.30 h and in the morning after sleep, subjective feelings of fatigue, activation and mood were assessed by a standardized adjective check list (Janke and Debus, 1978). Standard polysomnographical recordings were conducted between

23.00 and 7.00 h. To determine plasma concentrations of ACTH, cortisol, growth hormone (GH) and thyroid stimulating hormone (TSH) blood was collected every 30 min. Plasma concentrations of IFN-alpha, IL-6, IFN-gamma and of C-reactive protein were determined in samples collected at 18.30, 22.30, 1.30, and 6.30 h. During the wake periods prior to and after sleep body temperature (sublingual), blood pressure and heart rate were measured every hour.

The adjective check list applied before sleep indicated that subjects following the high dose of IFN-alpha felt more tired and less activated than following placebo ($p < 0.05$). Table 1 summarizes the effects of IFN-alpha on sleep. Both doses of IFN-alpha distinctly reduced time spent in SWS. A separate analysis of the first and second half of sleep revealed this suppressing effect of the cytokine on SWS to concentrate on the first half of the night (Fig. 2). The time spent in REM sleep was also suppressed after IFN-alpha. But, this effect reached significance for the high dose of IFN-alpha, only. Also, statistical significance for the suppression of REM sleep was revealed, only, when the entire night was analyzed. High doses of IFN-alpha, in addition, exerted an increasing influence on time spent in stage 1 sleep. Together, these changes indicate an acute disrupting effect of IFN-alpha on sleep which was more pronounced after the high than low dose administered.

The reduction in SWS following IFN-alpha concentrating during the first half of the night was accompanied by significantly enhanced plasma levels of cortisol, ACTH and GH (Fig. 3). Plasma levels of TSH tended to be decreased after IFN-alpha. However, the effect did not reach significance.

Plasma concentrations of IFN-alpha measured at 22.30 h as well as at the succeeding points in time, as expected, did not reveal any significant increase after administration of the cytokine. This was not unexpected since the estimated plasma levels after such low doses of exogenous IFN-alpha remain below the detection limit of the assays currently available. IFN-alpha at the dose of 10,000 IU/kg bodyweight increased concentrations of IL-6, measured at 1.30 h (mean +/–SEM: 10.8 +/– 4.2 versus 1.74 +/– 0.5 pg/ml on the placebo condition, $p < 0.01$) and at 6.30 h (3.0+/–0.6 versus 1.64 +/– 0.4, $p < 0.05$) which clearly indicates that the cytokine was biologically active even at the fairly low concentrations induced. Also, the high dose of IFN-alpha enhanced plasma concentrations of IFN-gamma at 6.30 h (24.3 +/– 8.4 versus 17.1 +/– 10.4 pg/ml, following placebo, $p < 0.05$). Measures of C-reactive protein remained unaffected.

At the doses administered, IFN-alpha did not induce fever, although in the morning following sleep temperature was slightly increased (by 0.37°C on average) following the high dose of IFN-alpha as compared with the effect of placebo ($p < 0.005$). Blood pressure and heart rate, in gernal were not affected by the cytokine.

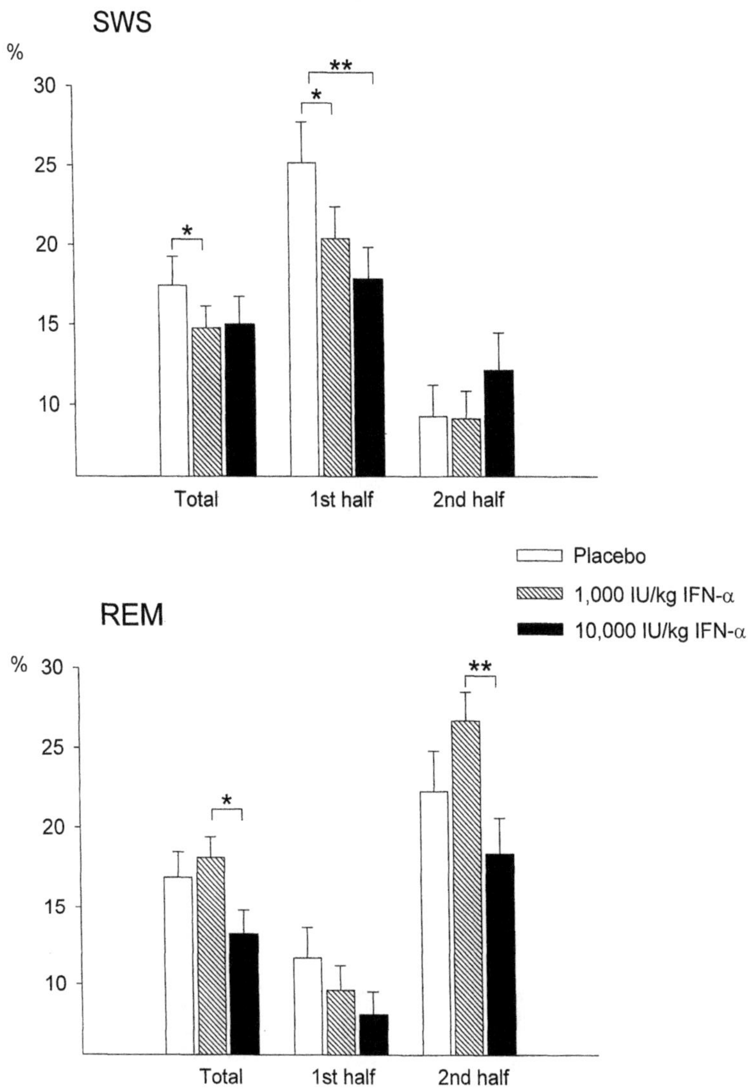

Fig. 2. Mean (+/– SEM) per cent of time spent in SWS (top) and in REM sleep (bottom) following administration of placebo (empty columns), a low dose of IFN-alpha (1000 IU/kg body weight, hatched columns) and a high dose of IFN-alpha (10,000 IU/kg/body weigth, black columns). Percentages were calculated for the entire nocturnal sleep period ("Total", left) as well as separately for the first (middle) and second half of the night (right). Substances were administered subcutaneously at 19.00 h prior to experimental nights. (n = 18). **–$p < 0.01$, *–$p < 0.05$, for pairwise comparisons between the effects of any two treatments (data are from Späth-Schwalbe et al., 1999)

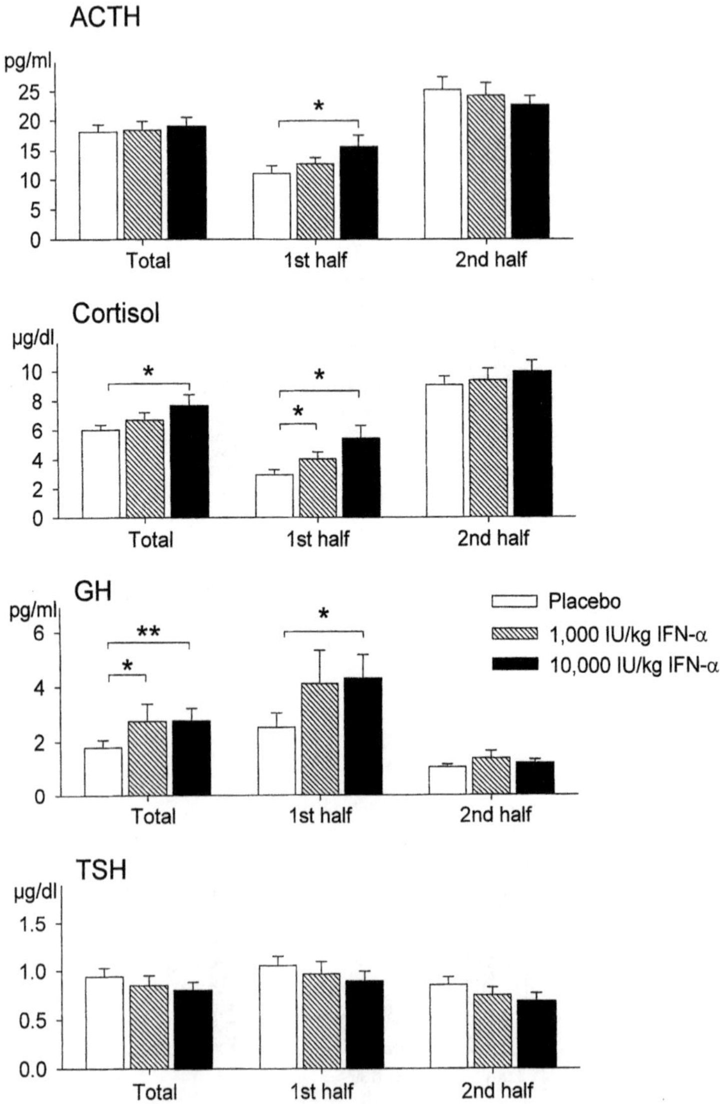

Fig. 3. Mean (+/– SEM) plasma concentrations of (from top to bottom) adrenocorticotropin (ACTH), cortisol, growth hormone (GH), and thyroid stimulating hormone (TSH) following administration of placebo (empty columns), a low dose of IFN-alpha (1000 IU/kg body weight, hatched columns) and a high dose of IFN-alpha (10,000 IU/kg/body weigth, black columns). Average concentrations were calculated for the entire nocturnal sleep period ("Total", left) as well as separately for the first (middle) and second half of the night (right). Substances were administered subcutaneously at 19.00 h prior to experimental nights. (n = 18). **–$p < 0.01$, *–$p < 0.05$, for pairwise comparisons between the effects of any two treatments (data are from Späth-Schwalbe et al., 1999)

Table 1. Sleep onset, total sleep time and absolute time (in min) spent in the different sleep stages and latency (in min) for each sleep stage after placebo and subcutaneous administration of IFN-α (1000 and 10,000 IU/kg body weight)

	Placebo (P)		1000 IU/kg IFN-α (L)		10,000 IU/kg IFN-α (H)		Significance P<		
	Mean	SEM	Mean	SEM	Mean	SEM	P–L	P–H	L–H
Sleep onset	16.7	4.2	21.3	6.8	34.2	11.7			
Sleep Time	445.7	7.3	441.0	9.5	428.2	12.9			
Wake	37.6	10.1	38.8	13.0	37.2	8.6			
S1	42.2	4.0	43.4	5.8	55.0	7.6		0.04	
S2	212.6	13.9	213.7	13.0	215.2	13.8			
SWS	77.7	7.6	65.1	5.5	64.2	6.5	0.03	0.07	
REM	75.6	7.4	80.1	6.4	56.7	7.6		0.03	0.002
Latency									
S2	5.53	0.7	5.60	0.9	4.44	0.5			
SWS	28.25	9.6	23.10	5.2	24.03	6.5			
REM	103.78	16.8	123.75	18.4	161.60	26.7		0.02	

S1, S2 sleep stage 1 and 2; *SWS* slow wave sleep, i.e. sleep stages 3 + 4; *REM* rapid eye movement sleep. Right columns indicate significant differences between the effects of any two of the treatments. *P* Placebo, *L* low dose of IFN-α, *H* high dose of IFN-α

Comparison of effects of IFN-alpha and IL-6

Since in a foregoing study (Späth-Schwalbe et al., 1998) the effects of IL-6 (0.5 μg/kg body weight, injected subcutaneously at 19.00 h) on nocturnal sleep in healthy men were examined under nearly identical conditions, a comparison of the influences of both cytokines is tempting. While IFN-alpha is a cytokine critically involved in the immune defense against viral infection, IL-6 seems to play an equally important role in the regulation of the immune defense against bacterial infectious challenge.

As depicted in Table 2, the profile of action of IL-6 in essential aspects bears striking similarities with that obtained for IFN-alpha. Both, IFN-alpha and IL-6 acutely reduced SWS with this effect concentrating during the first half of the night. Also, the time spent in REM sleep was diminished across the entire nocturnal period of sleep after the high dose of IFN-alpha as well as after IL-6. Self-report measures indicated that after both cytokines subjects felt less activated and more tired as compared to the effects of placebo. IL-6 in addition impaired mood in that subjects following IL-6 administration scored lower on the "high spirits" scale of the EWL adjective check list. With regard to endocrine measures, IFN-alpha like IL-6 stimulated the release of ACTH and cortisol from the pituitary-adrenal system. After both cytokines, the effect concentrated during the first half of the night.

Table 2. Comparison of effects of IL-6 and IFN-α in humans

	IL-6	IFN-α
Central nervous sleep:	SWS during early sleep ↓ REM sleep ↓	SWS during early sleep ↓ REM sleep ↓
Self-report:	Deactivation and fatigue ↑ High Spirits ↓	Deactivation and fatigue ↑
Endocrine system:	ACTH during early sleep ↑ Cortisol during early sleep ↑ GH – no effect TSH ↓	ACTH during early sleep ↑ Cortisol during early sleep ↑ GH during early sleep ↑ TSH – not significant
Immune system:	Heart rate ↑ Body temperature ↑ CRP ↑ IFN-γ – no effect	Heart rate – no effect Body temperature – no effect CRP – no effect IFN-γ ↑ IL-6 ↑

Effects after subcutaneous injection of IL-6 0.5 μg/kg body weight and of IFN-α 10,000 IU/kg body weight at 19.00 h prior to experimental sleep in young healthy humans

But, divergent actions were also revealed: Solely, IFN-alpha induced a pronounced increase in GH concentrations during the first half of the night. On the other hand, a decline in TSH levels revealed to be significant solely following IL-6. Reductions in TSH concentrations after IFN-alpha were weaker and failed to reach significance. In addition, IL-6 induced distinct increases in body temperature and heart rate whereas corresponding changes following IFN-alpha remained negligible. Moreover, after IL-6 a delayed increase in concentrations of C-reactive protein was obtained which did not occur after IFN-alpha. On the other hand, IFN-alpha administration induced an increase in IFN-gamma plasma concentrations which was not revealed after IL-6.

Concluding remarks

The acute changes in sleepiness and sleep and in the associated endocrine activity observed following administration of IFN-alpha shared remarkable similarities with those observed following IL-6, suggesting common mechanisms mediating the central nervous effects of both cytokines. In this respect, the stimulating effect on the release of IL-6 observed following the administration of IFN-alpha, points to a possible contribution of endogenous IL-6 to the effects of IFN-alpha. However, the increase in IL-6 concentrations after IFN-alpha did not develop immediatley after administration of the cytokine: i.e., the concentrations measured at 22.30 h (3.5 hours post-injection) were still comparable for the placebo and IFN-alpha conditions. Moreover, an enhancement in IL-6 concentrations was obtained after the high dose of IFN-alpha, only, although reduced SWS and increased concentrations of ACTH/cortisol were observed during the first half of nocturnal sleep likewise after the low dose of IFN-alpha. These temporal and pharmacodynamic features warn against an overestimation of the role of IL-6 in mediating the influences of IFN-alpha.

Both cytokines stimulated secretory activity of the pituitary-adrenal system which is consonant with foregoing findings in patients (Mastorakos et al., 1993; Späth-Schwalbe et al., 1989, 1994). Administration of cortisol and ACTH in humans has been shown to inhibit REM sleep (Born et al., 1989). Accordingly, the decreasing effect of the cytokines on REM sleep was most likely a consequence of their stimulating influence on this endocrine axis. Enhanced secretion of cortisol may also explain the decline in TSH concentration, since glucocorticoids are known to inhibit TSH release (Wilber and Utiger, 1969; Brabant et al., 1987). However, stimulation of pituitary-adrenal secretory activity cannot explain the acute suppressing influences of IFN-alpha and IL-6 on SWS. Neither administration of cortisol nor of ACTH, in previous studies in humans revealed a similar effect. Also, the increased subjective tiredness after cytokine administration can be hardly explained by increased release of cortisol and ACTH (Plihal et al., 1996; Born et al., 1990). A mediation of these effects by changes body temperature and heart rate is likewise excluded, since after IFN-alpha the feelings of tiredness and suppression of SWS developed without any significant concommitant changes in these parameters. These changes, therefore, probably involve different as yet unknown mechanisms. Since systemic IFN-alpha and IL-6 have been shown to enter the cerebrospinal fluid compartment (Smith et al., 1985; Banks et al., 1994), a direct action on central nervous structures involved in sleep regulation is conceivable.

Effects on sleep observed in healthy humans after the administration of IFN-alpha and IL-6 contrast with those observed in rats and rabbits. In rabbits, IL-6 did not affect sleep (Opp et al., 1989) whereas IFN-alpha in rats and rabbits – as described above – distinctly enhanced SWS. However, REM sleep was transiently diminished after IFN-alpha. In humans, an acutely disrupting effect of IFN-alpha as well as of IL-6 on sleep prevailed with, both the amount of SWS and of REM sleep being distinctly reduced. The effects on SWS concentrated during early sleep. It has been suggested in this context, that the sleep promoting effect of proinflammatory cytokines may be modulated by circadian oscillations and, hence, could depend on the time of day of cytokine administration. Effects on daytime sleep of cytokines in humans have not been examined so far. However, in previous studies, Pollmächer's group (Pollmächer et al., 1993; Korth et al., 1996) demonstrated that the administration of endotoxin mimicking a bacterial infectious challenge with distinct increases in the release of IL-6 and TNF-alpha, did not increase SWS in humans irrespectively of whether administered in the evening or in the morning. Therefore, a species specifity in the regulatory influence of IFN-alpha on central nervous sleep processes cannot be ruled out, and may pertain also to other proinflammatory cytokines.

Interestingly, in several aspects the central nervous effects observed after IFN-alpha and IL-6 are reminescent of symptoms that can be observed in depressed patients who have been found also to display increased plasma concentrations of IL-6 (Maes et al., 1995, see also Maes, this volume). Depressed patients typically complain of feelings of deactivation and subjective fatigue. But, at the same time, their sleep is disturbed showing reduced SWS and a dysregulation of REM sleep (e.g., Benca et al., 1992; Kupfer and Reynolds, 1992). Hyperactivity of pituitary-adrenal secretion belongs to a major symptom in these patients (Halbreich et al., 1985; Barden et al., 1995). In addition, depression can be associated with diminished TSH release (Mason et al., 1995). Following IL-6, healthy subjects also displayed signs of impaired mood as evaluated by self-report. In light of these parallels a contribution of host defense mechanisms to the disease has to be taken into consideration.

Acknowledgements

This work was supported by the Volkswagen-Stiftung (E.S.S.) and the Deutsche Forschungsgemeinschaft (J.B.). We thank A. Otterbein for helpful assistance with preparing the manuscript.

References

Banks WA, Kastin AJ, Gutierrez EZ (1994) Penetration of interleukin-6 across the murine blood-brain barrier. Neurosci Lett 179:53–56

Benca RM, Obermeyer WH, Thisted RA, Gillin JC (1992) Sleep and psychiatric disorders: a meta-analysis. Arch Gen Psychiatry 49:651–668

Born J, Späth-Schwalbe E (1997) Effects of cytokines on human EEG and sleep. In: Wieselmann G (ed) Current update in psychoimmunology. Springer, Wien New York, pp 103–118

Born J, Späth-Schwalbe E, Schwakenhofer H, Kern W, Fehm HL (1989) Influences of corticotropin-releasing hormone, adrenocorticotropin, and cortisol on sleep in normal man. J Clin Endocrinol Metab 68:904–911

Born J, Bathelt B, Pietrowsky R, Pauschinger P, Fehm HL (1990) Influences of peripheral adrenocorticotropin 1–39 (ACTH) and human corticotropin releasing hormone (h-CRH) on human auditory evoked potentials (AEP). Psychopharmacol 101:34–38

Brabant G, Brabant A, Ranft U, Ocran K, Köhrle J, Hesch RD, von zur Mühlen A (1987) Circadian and pulsatile thyrotropin secretion in euthyroid man

under the influence of thyroid hormone and glucocorticoid administration. J Clin Endocrinol Metab 65:83–88
Barden N, Reul JMHM, Holsboer F (1995) Do antidepressants stabilize mood through actions on the hypothalamic-pituitary-adrenal system? Trends Neurosci 18:6–11
DeMaeyer E, DeMaeyer-Guignard J (1994) Interferons. In: Thomson A (ed) The cytokine handbook. Academic Press, London, pp 265–288
DeSarro GB, Asciotti C, Audino MG, Rispoli U, Nistico G (1989) Behavioral and ECOG spectrum changes induced by intracerebral microfusion of interferons and interleukin-2 in rats are antagonized by naloxone. In: Hadden JW, Masek K, Nistico G (eds) Interactions among central nervous system, neuroendocrine and immune systems. Pythagora Press, Rome, pp 351–371
Guilleminault C, Mondini S (1986) Mononucleosis and chronic daytime sleepiness: a longterm follow-up study. Arch Intern Med 146:1333–1335
Habif DV, Lipton R, Cantell K (1975) Interferon crosses blood-cerebrospinal fluid barriert in monkeys. Proc Soc Exp Biol Med 149:287–289
Halbreich U, Asnis GM, Shindledecker R, Zumoff B, Nathan S (1985) Cortisol secretion in endogenous depressinon. II. Time-related functions. Arch Gen Psychiatry 42:909–914
Janicki PK (1992) Binding of human alpha-interferon in the brain tissue membranes of rat. Res Comm Chem Pathol Pharmacol 75:117–119
Janke W, Debus G (1978) Die Eigenschaftswörterliste EWL. Eine mehrdimensionale Methode zur Beschreibung von Aspekten des Befindens. Verlag für Psychologie-Hogrefe, Göttingen
Kawakami M, Negoro H, Terasawa E (1965) Influence of immobilization stress upon the paradoxical sleep in the rabbit. Jph J Physiol 15:1–11
Kimura M, Majde JA, Toth LA, Opp MR, Krueger JM (1994) Somnogenic effects of rabbit and human recombinant interfrerons in rabbits. Am J Physiol 267:R53
Korth C, Mullington J, Schreiber W, Pollmächer T (1996) Influence of endotoxin on daytime sleep in humans. Infect Immun 64:1110–1115
Krüger JM, Majde JA (1994) Microbial products and cytokines in sleep and fever regulation. Crit Rev Immunol 14:355–379
Krüger JM, Dinarello CA, Shoham S, Davenne D, Walter J, Kubillus S (1987) Interferon alpha-2 enhances slow-wave sleep in rabbits. Int J Immunopharmacol 9:23–20
Kubicki S, Henkes H, Terstegge K, Ruf B (1988) AIDS related sleep disturbances – a preliminary report. In: Kubicki S, Henkes H, Bienzle U, Pohle U (eds) HIV and nervous system. Gustav Fischer, New York, pp 97–105
Kupfer DJ, Reynolds CF (1992) Sleep and affective disorders. In: Paykel ES (ed) Handbook of affective disorders. Churchill Livingstone, New York, pp 311–323
Mason GA, Garbutt JC, Prange AJ (1995) Thyrotropin-releasing hormnoe. In: Bloom FE, Kupfer DJ (eds) Psychopharmacology – the fourth generation of progress. Raven Press, New York, pp 493–504
Mastorakos G, Chrousos GP, Weber JS (1993) Recombinant interleukin-6 activates the hypothalamic-pituitary-adrenal axis in humans. J Clin Endocrinol Metab 77:1690–1694
Maes M, Meltzer HY, Bosmans E, Bergmans R, Vandoolaeghe E, Ranjan R, Desnyder R (1995) Increased plasma concentrations of interleukin-6,

soluble interleukin-6, soluble interleukin-2 and transferrin receptor in major depression. J Affect Disord 34:301–309

Moldofsky H (1995) Sleep, neuroimmune and neuroendocrine functions in fibromyalgia and chronic fatigue syndrome. Adv Neuroimmunol 5:39–56

Norman SE, Chediak HD, Kiel M, Cohn MA (1990) Sleep disturbances in HIV-infected homosexual men. AIDS 4:775–781

Norman SE, Chediak HD, Freeman C, Kiel M, Mendez A, Simoneau J, Nolan B (1992) Sleep disturbance in men with asymptomatic human immunodeficiency infection. Sleep 15:150–155

Opp M, Obal F Jr, Cady AB, Johannsen L, Krüger JM (1989) Interleukin-6 is pyrogenic but not somnogenic. Physiol Behav 45:1069–1072

Plihal W, Krug R, Pietrowsky R, Fehm HL, Born J (1996) Corticosteroid receptor mediated effects on mood in humans. Psychoneuroendocrinology 21:513–523

Pollmächer T, Schreiber W, Gudewill S, Trachsel L, Galanos C, Holsboer F (1993) Influence of endotoxin on noctrunaol sleep in humans. Am J Physiol 264:R1077–R1083

Reite M, Laudenslager M, Jones J, Crnic L, Kaemingk K (1987) Interferon decreases REM latency. Biol Psychiat 22:104–107

Smedley H, Katrak M, Sikora K, Wheeler T (1983) Neurological effects of recombinant human interferon. Br Med J 286:262–264

Smith RA, Norris F, Palmer D, Bernhardt L, Wills RJ (1985) Distribution of alpha interferon in serum and cerebrospinal fluid after systemic administration. Clin Pharmacol Ther 37:85–88

Späth-Schwalbe E, Porzsolt F, Kigel W, Born J, Kloss B, Fehm HL (1989) Elevated plasma cortisol levels during interferon-γ treatment. Immunopharmacology 17:141–145

Späth-Schwalbe E, Born J, Schrezenmeier H, Bornstein SR, Stromeyer P, Drechsler S, Fehm HL, Porzsolt F (1994) Interleukin-6 stimulates the hypothalamus-pituitary-adrenocortical axis in man. J Clin Endocrinol Metab 79:1212–1214

Späth-Schwalbe E, Hansen K, Schmidt F, Schrezenmeier H, Marshall L, Burger K, Fehm HL, Born J (1998) Acute effects of recombinant human interleukin-6 on endocrine and centralnervous sleep functions in healthy men. J Clin Endocrinol Metab 83:1573–1579

Späth-Schwalbe E, Perras B, Fehm HL, Born J (1999) Acute effects of recombinant interferon-α on nocturnal sleep in healthy men submitted

Von Economo C (1930) Sleep as a problem of localization. J Nerv Ment Dis 71:249–259

Wilber JF, Utiger RD (1969) The effect of glucocorticoids on thyrotropin secretion. J Clin Invest 48:2096–2103

Correspondence: Prof. Dr. J. Born, Klinische Forschergruppe – Klinische Neuroendokrinologie, Medizinische Universität Lübeck, Ratzeburger Allee 160, Haus 23a, D-23538 Lübeck, Federal Republic of Germany

Immunological alterations in patients suffering from bulimia nervosa (DSM-IV 307.51) vs anxiety disorders (DSM-IV 300.01, 300.21, 309.81, 300.02)

G. Wieselmann[1], **G. Herzog**[1], **M. Schönauer-Cejpek**[1], **U. Demel**[2], **K. Penkner**[3], **G. Tilz**[2], and **H.G. Zapotoczky**[1]

Departments of [1]Psychiatry, [2]Clinical Immunology, and [3]Prosthetic Dentistry, University Hospital, Graz, Austria

Summary

Alterations of the immunesystem in different kinds of psychiatric disorders are well known in literature in the last two decades. While there is a remarkable impact of these findings for patients suffering from "endogenous psychoses", the role of the immunesystem in former called "neuroses" is still not clarified. Previous attempts to evaluate these ideas showed a relationship between anxiety disorders and alterations of the immunesystem.

The aim of this pilot study was to investigate the possible role of the immunesystem (cellular activity, humoral/polyclonal activation) in patients suffering from bulimia nervosa (DSM-IV 307.51) and to compare these data to those of sex and age matched anxiety disorder patients.

The significance of immunological alterations in both groups of mental disorders was verified but the causes for these alterations remain unclear and need further research.

Introduction

Alterations of the immunesystem in psychiatric disorders are well known in the literature. While there is a remarkable impact of these findings for patients suffering from "endogenous psychoses" (Wieselmann et al., 1997), the role of the immunesystem in so called "neuroses" is still not clarified but there is a high incidence of psychoimmunological alterations in patients suffering from anxiety disorders (Surmann et al., 1986; Stein et al., 1988).

Pomeroy et al. (1994) investigated the role of interleukin-6, tumor necrosis factor-alpha and transforming-growth-factor-beta in anorexia nervosa. Both were significantly elevated during starvation and returned to levels comparable to those of normal-weight controls after weight gain. The aim of this pilot study was to investigate immunological alterations in patients suffering from bulimia nervosa and to compare these data with those of sex and age matched anxiety patients.

Methods and materials

23 females suffering from bulimia nervosa (BN) were compared to 23 sex and age matched anxiety disorder patients (AD) (mean age 22.98, sdev. 3.36, range from 17 to 31).

BN-patients

All patients were sent to the outpatient unit and underwent the following *diagnostic clarification:*

- patient's history
- Three factors Eating Questionnaire (TFEQ, germ. vers.)
- physical examination
- ECG, EEG
- routine laboratory test
- bTSH, fT3, fT4
- clinical examination of the Temporo-Mandibular-Joint (TMJ)-system
- immuneseraprofile (to evaluate the cellular activity as well as the humoral/polyclonal activation):

soluble-interleukin-2-receptors (*s-IL-2R*, 529–913 U/ml), *cardiolipin antibodies* (max. 12 U/ml), beta2-microglobulin (*Beta2M* max. 2,4 mg/l), *T8* (138–533 U/ml), intracellular adhesion molecules (*I-CAM* 129,9–297μg/ml), gammaglobulin synthesis (*IgG* 6,8–15,3 g/l; *IgA* .74–3.74 g/l; *IgM* .40–2.48 g/l), circulating immunecomplexes (*CIC* max. 4 μEq/ml), complementfactors *C3* (.75–1.35 g/l), *C4* (.18–.45 g/l), auto-antibodies to nuclear factors (*ANF* pos/neg), auto-antibodies to *ds-DNA* (pos/neg), auto-antibodies to glomerulobasementmembrane (*GBM*, pos/neg), auto-antibodies to basementmembrane *type I* and *-type II* (pos/neg.) and anti-neutrophile-cytoplasma-antibodies (*ANCA* pos/neg). These specific and nonspecific parameters were quantitated by nephelometry and indirect immunofluorescence.

AD-patients

We selected 23 age matched females, according to DSM-IV criteria for "anxiety disorder": 8 panic, 9 panic with agoraphobia, 2 PTSD and 4 GAD. All patients were drugnaive.

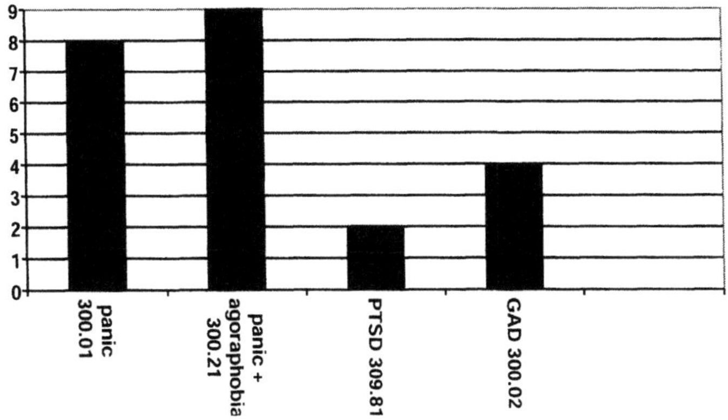

Fig. 1. Number of the DSM-IV criteria for "anxiety disorder"

Diagnostic clarification:
- patient's history
- SCID-Interview
- physical examination
- ECG, EEG
- routine laboratory test
- bTSH, fT3, fT4
- drug urine screening
- clinical examination of the TMJ-system
- immuneseraprofile

Results

Because of the main objective of this study the authors only will deal with the following findings:

Description of the Body-Mass-Index (BMI)

– mean BMI (Body-Mass-Index) = 20.09 / median = 19.72 / sdev. 2.4 / from 16.3 to 26.8 (two patients scored slightly under 17.5)

All BN-patients suffered from recurrent episodes of binge eating associated with selfinduced vomiting (median = 7.5 per week). The mean duration was 80.8 months (median = 62 months) (DSM-IV 307.51). One patient misused sometimes appetite suppressants, five patients seldom misused laxatives.

BN-patients with immunological alterations did not differ from BN-patients without immunological alterations in any of the parameters as shown before.

Drug-treatment

Eight patients were absolutely drugnaive, the other bulimics were former treated by general practitioners and/or psychiatrists with fluoxetine and/or thioridazine, dixyrazine or amytriptiline, one patient was additionally on naltrexone. After a washout-period of 1 week we took the bloodsample of these patients and did all the other investigationes as mentioned above.

Three-Factor Eating Questionnaire (TFEQ, Stunkard and Messick, 1985, germ. Vers. Pudel and Westenhöfer, 1989)

Results of TFEQ (Table 1):

Table 1. Results of three-factor eating questionnaire

	Mean	Std Dev
TFEQ-cognitive restraint scale	12,7778	4,2086
TFEQ-disinhibition scale	12,3889	3,4324
TFEQ-hunger scale	9,8889	3,4792

TFEQ-cognitive restraint scale (Mean = 12,7778; sdev = 4,2086)
TFEQ-disinhibition scale (Mean = 12,3889; sdev = 3,4324)
TFEQ-hunger scale (Mean = 9,8889; sdev = 3,4792)
Concerning these three scores there are no differences between BN-patients with or without immunealterations.

Alterations of the immunesystem

Abnormal immune profiles are rather frequent in both groups (BN 13 normal, 6 slightly abnormal and 4 pathological) and immunological alterations are more common in the anxiety group (AD 8 normal, 10 slightly abnormal and 5 pathological), (Table 2) mainly in the

Table 2. Immunealterations in different patient groups

	Normal	Slightly abnormal	Abnormal	Total
Bulimia nervosa	13	6	4	23
Anxiety disorder	8	10	5	23
	21	16	9	46

$\chi^2 = 2{,}30159$; df = 2 p = .31 n. s.

Table 3. Immunological parameter differences between the two patient groups

Parameter	Mean bulimic / sdev.	Mean anxiety / sdev.	t	df	p	Number of patients with the path. parameters – bulimia ↑ increased ↓ decreased	Number of patients with the path. parameters – anxiety ↑ increased ↓ decreased
IgG	10,5357 / 2,294	15,2539 / 2,567	6,57	44	,000	1 ↑	**12** ↑
IgA	1,8191 / ,537	2,3283 / ,766	2,61	39,42	,013	0	0
IgM	1,5987 / ,494	1,9964 / ,864	1,92	35,01	,063	0	**7** ↑
Beta2M	1,2955 / ,268	1,5435 / ,477	2,16	34,94	,038	0	1 ↑
CIC C1q	2,6741 / 2,391	1,6232 / 1,419	1,77	34,16	,085	**5** ↑	**3** ↑

part of slightly abnormal immune parameters. However statistical significant levels could not be reached (χ^2 =2,30159; df = 2 p = .31 n. s).

The analysis of the sera immune parameters showed differences between the patients groups (Table 3). Both groups showed increased CIC-levels, but there was an increase of gammaglobuline synthesis in the group of the AD-patients only.

No significant differences were found for C3, C4, T8, I-CAM, ds-DNA, sIL-2r. No abnormal alterations could be seen in both groups concerning: antibodies to cardiolipin, ANCA, auto-antibodies to basementmembrane-type II.

One BN-patient showed auto-antibodies to GBM, one AD-patient showed auto-antibodies to basementmembrane-type I. Some positive ANF alterations were found. However, these results did not reach statistical significance.

Discussion

The analysis of this pilotstudy showed abnormal immune-profiles of different combinations in BN-patients as well as in AD-patients. Increased levels of CIC had the highest prevalence of immunological alterations in both groups, whereas elevated gammaglobuline-synthesis is much more specific for AD-patients. These results are in accordance to previous findings in patients suffering from anxiety disorders (Langs et al., 1997).

The question arises if there are specific immunological markers for different mental disorders. However, the significance of immunological alterations in mental disorders was verified again in this study; but the causes for this alterations remain unclear and need further investigations.

References

American Psychiatric Association (1994) DSM IV diagnostic and statistical manual for mental disorders, 4th ed. American Psychiatric Press, Washington, D.C.

Langs G, Herzog G, Penkner K, Demel U, Tilz G, Bratschko RO, Wieselmann G (1997) Psychoimmunology, anxiety disorders and TMJ-disorders. In: Wieselmann G (ed) Current update in psychoimmunology. Springer, Wien New York, pp 79–83 (Key Topics in Brain Research)

Pomeroy C, Eckert E, Hu S, Eiken B, Mentink M, Crosby RD, Chao CC (1994) Role of interleukin-6 and transforming growth factor-beta in anorexia nervosa. Biol Psychiatry 36:836–839

Stein M, Keller SE, Schleifer SJ (1988) Immunesystem: relationship to anxiety disorders. Psychiatr Clin North Am 11:349–360

Surmann OS, Williams J, Sheehan DV, Strom TB, Jones KJ, Coleman J (1986) Immunological response to stress in agoraphobia and panic attacks. Biol Psychiatry 21:768–774

Stunkard AJ, Messick S (1985) The three-factor eating questionnaire to measure dietary restraint, disinhibition and hunger. J Psych Res 29:71–83

Wieselmann G, Tilz G, Fabisch H, Herzog G, Langs G, Demel U, Zapotoczky HG (1997) Immunological factors in the sera of patients with acute and chronic schizophrenia. In: Henneberg AE, Kaschka WP (eds) Immunological alterations in psychiatric diseases. Karger, Basel, pp 85–89 (Adv Biol Psychiatry 18)

Correspondence: Univ.-Prof. Dr. G. Wieselmann, Universitätsklinik für Psychiatrie, Auenbruggerplatz 22, A-8036 Graz, Austria

T cell subsets and expression of integrins in peripheral blood of patients with migraine and chronic tension type headache

M. Empl[1], **P. Sostak**[1], **M. Breckner**[2], **M. Riedel**[2], **N. Müller**[2], **R. Gruber**[3], **S. Förderreuther**[1], and **A. Straube**[1]

Departments of [1]Neurology, [2]Psychiatry, and [3]Immunology, Ludwig-Maximilians-University, Munich, Federal Republic of Germany

Summary

Neurogenic inflammation of dural vessels has been hypothesised to be an essential part of migraine physiopathology. Analysis of lymphocyte subpopulations and integrin expression in the peripheral blood of headache patients (migraine and chronic tension type headache) showed an elevated proportion of T-helper cells and T-helper memory cells as well as a higher expression of LFA-1 on T-helper and T-suppressor cells. This could be a sign of enhanced endothelium and lymphocyte interaction as it would be expected in an inflammatory process of cranial vessels.

Introduction

Interactions between the nervous and the immune system – well-known in the mediation of inflammatory pain – have been hypothesised to be an essential part of the pathophysiology of migraine, following the animal model of neurogenic inflammation. In animals, antidromic trigeminal stimulation causes the release of sensory neuropeptides as substance P, neurokinin A and calcitonin-gene related peptide, leading to endothelial activation, plasmaextravasation and platelet activation in the postcapillary venules of the dura mater (Dimitriadou et al., 1991), effects, that could be blocked by migraine drugs such as sumatriptan, ergotamin or aspirin (Buzzi et al., 1992, 1989; Goadsby and Edvinsson, 1993). In humans, an elevation of CGRP in the blood of the jugular vein could be shown during migraine attacks (Goadsby et al., 1990), with a normalization subsequent to the administration of sumatriptan and the relief of headache (Goadsby and Edvinsson, 1993).

However, despite the fact, that drugs resolving acute migraine attacks block neurogenic inflammation in the animal model, the existence of neurogenic inflammation as an essential part in the human pathophysiology of migraine remains to be demonstrated. Otherwise, the efficacy of anti-inflammatory analgesics in different types of headaches, suggests that neurogenic inflammation may even serve as a model for the pathophysiology of other primary headaches.

Endothelium and T-cell interaction in neurogenic inflammation

Since an inflammatory process of blood vessels is accompanied by endothelial activation, as shown in the animal model, and endothelial activation leads to enhanced adhesion of T cells to the veinous vascular endothelium (Roit et al., 1998), markers of enhanced interaction like the expression of integrins on lymphocytes (LFA-1 and VLA-4), that are essential for lymphocyte adhesion to the endothelium (Roit et al., 1998) and possibly contribute to T-cell activation (Lub et al., 1995; van Schooten and Quan, 1993), should be observed.

To evaluate the possible interaction of lymphocytes with the cerebral endothelium, we analyzed the expression of integrins on lymphocytes (LFA-1 and VLA-4) as well as the proportion of lymphocyte subsets ($CD3^+$, $CD4^+$, $CD8^+$, and $CD45RO^+$) as a possible indicator for immunological activation in the peripheral blood of 20 patients with primary headache (migraine with and without aura, chronic tension type headache, as defined by the IHS-criteria) by trichromic flow cytometry. The results were compared with those of 16 headache-free controls.

Patients with chronic headache, during attacks or in attack-free periods, showed higher proportions of $CD4^+$ T-helper cells ($46.9 \times 10\%$, vs. $40.6 \times 4.9\%$, $p \leq 0.006$), a higher proportion of $CD4^+45RO^+$ memory cells ($23.5 \times 5.3\%$ vs. $16.7 \times 5.8\%$, $p \leq 0.001$), and a significantly higher expression of LFA-1 on $CD4^+$ cells ($30.1 \times 6.8\%$ vs. $23.0 \times 4.5\%$, $p \leq 0.002$) and on $CD8^+$ cells ($49.6 \times 13.6\%$ vs. $29.8 \times 5.0\%$, $p \leq 0.005$). In the expression of VLA-4 on $CD4^+$ or $CD8^+$ cells as well as in the proportion of $CD3^+$, $CD8^+$ and $CD8^+CD45RO^+$ cells, no differences could be found.

Possible significance of the enhanced endothelial-T-cell-interaction in the peripheral blood

The increased expression of LFA-1 on $CD4^+$ lymphocytes as well as the higher proportion of $CD4^+$ helper and $CD4^+CD45RO^+$ helper memory cells as a sign of previous stimulation suggests an immunological activation, as the expression of LFA-1 (occurring between 48 and 72 hours following an immunological stimulus) is upregulated following cell activation (Vermot Desroches et al., 1991) and is a prerequisite for the endothelial-T cell interaction. For the enhanced expression of LFA-1 on $CD4^+$ cells even in the headache free periods may account, that once strongly expressed after stimulation in vitro, LFA-1 expression remains on a relatively high level even after 10 days (Pryce et al., 1997), a quite normal interictal delay for patients consulting a headache clinic. Otherwise, the expression of VLA-4 is downregulated much faster (Pryce et al., 1997), which may help to explain why we did not see elevated VLA-4 levels. The interpretation of upregulated LFA-1 expression on lymphocytes may be limited by the missing determination of the activation state of LFA-1 by FACS-analysis, as the conformation of LFA-1 is essential for the functional role in adhesion and activation (Lub et al., 1995).

Furthermore, the finding that the expression of VLA-4 (very late antigen-4) is unchanged, may reflect the fact that the expression of VLA-4 in vivo is upregulated locally (p.e. in rheumatoid arthritis in the synovial liquid), but this local expression is not noticeable in the peripheral blood (Laffón et al., 1991).

As the enhanced expression of LFA-1 and higher proportion of $CD4^+$ and $CD4^+45RO$ memory cells were also observed in patients with chronic tension type headache, these preliminary results indicate a possible contribution of inflammatory reactions also to primary headache syndromes other than migraine. Further investigations of the pathophysiology of headaches, considering the possibilities of suppression of inflammatory reactions, may open new therapeutical options for headaches.

References

Buzzi MG, Sakas DE, Moskowitz MA (1989) Indomethacin and acetylic acid block plasma protein extravasation in rat dura mater following electrical stimulation of the rat trigeminal ganglia. Eur J Pharmacol 165:151–258

Buzzi MG, Dimitriadou V, Theoharides TC, Moskowitz MA (1992) 5-Hydroxytryptamine receptor agonist for the abortive treatment of vascular headaches block mast cell, endothelial and platelet activation within the rat dura mater after trigeminal stimulation. Brain Res 583:137–149

Dimitriadou V, Buzzi MG, Theoharides TC, Moskowitz MA (1991) Ultrastructural evidence for neurogenically mediated changes in blood vessels of the rat dura mater and tongue following antidromic trigeminal stimulation. Neuroscience 44:97–112

Goadsby PJ, Edvinsson L (1993) The trigeminovascular system and migraine: studies characterizing cerebrovascular and neuropeptide changes seen in humans and cats. Ann Neurol 33:48–56

Goadsby PJ, Edvinsson L, Ekman R (1990) Vasoactive peptide release in the extracerebral circulation of humans during migraine headache. Ann Neurol 29:183–187

Laffón A, García-Vicuña R, Humbría A, Postigo AA, Corbí AL, de Landázuri MO, Sánchez-Madrid F (1991) Upregulated expression and function of VLA-4 fibronectin receptors on human activated T cells in rheumatoid arthritis. J Clin Invest 88:546–552

Lub M, van Kooyk Y, Figdor CG (1995) Ins and outs of LFA-1. Immunol Today 16:479–483

Pryce G, Male D, Campbell I, Greenwood J (1997) Factors controlling T-cell migration across rat cerebral endothelium in vitro. J Neuroimmunol 75:84–94

Roitt I, Brostoff J, Male D (1998) Immunology, 5th ed. Mosby, London

Van Schooten WC, Quan J (1993) The influence of LFA-1 on human T cells stimulated by solid phase immobilized HLA class II-peptide complexes. Scand J Immunol 38:375–80

Vermot Desroches C, Rigal D, Andréoni C (1991) Regulation and functional involvement of distinct determinants of leucocyte Function-associated Antigen 1 (LFA-1) in T-cell activation in vitro. Scand J Immunol 33:277–286

Correspondence: Dr. M. Empl, Department of Neurology, University of Munich, Marchioninistrasse 15, D-81377 Munich, Federal Republic of Germany

Springer-Verlag
and the Environment

GPSR Compliance

The European Union's (EU) General Product Safety Regulation (GPSR) is a set of rules that requires consumer products to be safe and our obligations to ensure this.

If you have any concerns about our products, you can contact us on ProductSafety@springernature.com

In case Publisher is established outside the EU, the EU authorized representative is:

Springer Nature Customer Service Center GmbH
Europaplatz 3
69115 Heidelberg, Germany

Batch number: 10577896

Printed by Printforce, the Netherlands